The Open University

M249 Practical modern statistics

Book 1

Medical statistics

About this course

M249 Practical Modern Statistics uses the software packages *SPSS for Windows* (SPSS Inc.) and *WinBUGS*, and other software. This software is provided as part of the course, and its use is covered in the *Introduction to statistical modelling* and in the four computer books associated with *Books 1* to *4*.

Cover image courtesy of NASA. This photograph, acquired by the ASTER instrument on NASA's Terra satellite, shows an aerial view of a large alluvial fan between the Kunlun and Altun mountains in China's Xinjiang province. For more information, see NASA's Earth Observatory website at http://earthobservatory.nasa.gov.

This publication forms part of an Open University course. Details of this and other Open University courses can be obtained from the Student Registration and Enquiry Service, The Open University, PO Box 197, Milton Keynes, MK7 6BJ, United Kingdom: tel. +44 (0)870 333 4340, e-mail general-enquiries@open.ac.uk

Alternatively, you may visit the Open University website at http://www.open.ac.uk where you can learn more about the wide range of courses and packs offered at all levels by The Open University.

To purchase a selection of Open University course materials, visit the webshop at www.ouw.co.uk, or contact Open University Worldwide, Michael Young Building, Walton Hall, Milton Keynes, MK7 6AA, United Kingdom, for a brochure: tel. +44 (0)1908 858785, fax +44 (0)1908 858787, e-mail ouwenq@open.ac.uk

The Open University, Walton Hall, Milton Keynes, MK7 6AA.

First published 2007.

Copyright © 2007 The Open University

All rights reserved; no part of this publication may be reproduced, stored in a retrieval system, transmitted or utilised in any form or by any means, electronic, mechanical, photocopying, recording or otherwise, without written permission from the publisher or a licence from the Copyright Licensing Agency Ltd. Details of such licences (for reprographic reproduction) may be obtained from the Copyright Licensing Agency Ltd, Saffron House, 6–10 Kirby Street, London EC1N 8TS; website http://www.cla.co.uk.

Open University course materials may also be made available in electronic formats for use by students of the University. All rights, including copyright and related rights and database rights, in electronic course materials and their contents are owned by or licensed to The Open University, or otherwise used by The Open University as permitted by applicable law.

In using electronic course materials and their contents you agree that your use will be solely for the purposes of following an Open University course of study or otherwise as licensed by The Open University or its assigns.

Except as permitted above you undertake not to copy, store in any medium (including electronic storage or use in a website), distribute, transmit or re-transmit, broadcast, modify or show in public such electronic materials in whole or in part without the prior written consent of The Open University or in accordance with the Copyright, Designs and Patents Act 1988.

Edited, designed and typeset by The Open University, using the Open University TeX System.

Printed in the United Kingdom by Charlesworth Press, Wakefield.

ISBN 978 0 7492 1366 4

Contents

Study guide	5
Introduction	6
Part I Cohort studies and case-control studies	**7**
Introduction to Part I	7
1 Cohort studies	**8**
1.1 What is a cohort study?	8
1.2 Measures of association	11
1.3 Which measure of association should be used?	15
2 Models for cohort studies	**16**
2.1 The binomial model	17
2.2 Confidence intervals for the relative risk	18
2.3 Confidence intervals for the odds ratio	21
3 Case-control studies	**23**
3.1 What is a case-control study?	23
3.2 Measures of association in case-control studies	25
3.3 Studies with more than two exposure categories	29
4 Testing for no association in cohort studies and case-control studies	**31**
4.1 The chi-squared test statistic	31
4.2 The chi-squared test for no association	35
4.3 Fisher's exact test	40
5 Analysing cohort studies and case-control studies in SPSS	**41**
Part II Bias, confounding and causation	**42**
Introduction to Part II	42
6 Bias and confounding	**42**
6.1 What is bias?	43
6.2 Selection bias	43
6.3 Information bias	46
6.4 Confounding	47
7 Stratified analyses	**53**
7.1 The Mantel–Haenszel odds ratio	53
7.2 Matching in case-control studies	57
7.3 Interactions	62
8 From association to causation	**66**
8.1 Bradford Hill's criteria for causation	66
8.2 Dose-response analysis	67
9 Analysis of stratified tables in SPSS	**71**

Part III Randomized controlled trials and the medical literature — 72
Introduction to Part III — 72

10 Randomized controlled trials — 73
10.1 Randomization — 73
10.2 Concealment — 76
10.3 Analysis of randomized controlled trials — 78
10.4 Evaluation of pharmaceutical drugs — 81

11 Choosing the sample size — 85
11.1 Trial hypotheses, significance level and power — 85
11.2 An expression for the sample size — 88
11.3 Choosing the sample size — 90

12 Combining evidence from several studies — 94
12.1 Systematic reviews and meta-analysis — 94
12.2 Forest plots — 96

13 The medical literature — 100
13.1 The high altitude mountain sickness trial — 100
13.2 The Abstract and Introduction — 105
13.3 The Methods section — 107
13.4 The Results section — 108
13.5 The Discussion — 110

14 Exercises on Book 1 — 111

Summary of Book 1 — 115
Learning outcomes — 116

Solutions to Activities — 117

Solutions to Exercises — 127

Index — 132

Study guide

Each section of this book depends on ideas and results from the preceding sections, so we recommend that you study the sections in sequential order. The sections are of varying length — Section 7 will probably require the most study time, whereas Sections 8 and 12 are quite short.

This book contains both *activities*, which are included at various points in the text, and *exercises*, which are placed at the end of some sections. Their purposes are quite different. Activities form a central part of the text and you should try to do them as you work through the book. Exercises are provided to give you further practice at applying certain techniques and ideas *if you need it*: you should not routinely try them all as you study the book. You may find it more helpful to try them only if you are unsure that you have understood an idea. Or you may like to use them later in the year when you are revising for the examination. You will find a few further exercises in Section 14; some of these exercises cover material from more than one section. Solutions for most of the activities and exercises may be found at the back of this book; occasionally, a solution or a comment on an activity is included in the text following the activity. Activities for which you need to use your computer are contained in *Computer Book 1*. A few computer exercises are included at the end of the computer book. The computer book solutions are organized in a similar way to this book.

In Section 13 you will work through an article which has been reprinted in Subsection 13.1. You might find it convenient to have a loose copy to refer to as you work through Section 13. The article is available on the M249 course website, and you may print an extra copy from there if you wish.

As you study this book you will be asked to work through the five chapters of *Computer Book 1*. We recommend that you work through them at the points indicated in the text — Chapters 1 and 2 in Section 5, and Chapters 3, 4 and 5 in Section 9.

You should schedule fifteen study sessions for this book. This includes time for working through *Computer Book 1*, answering the TMA questions and consolidating your work on this book. You should schedule six study sessions for Part I, four for Part II, and five for Part III.

One possible study pattern is as follows.

Part I

Study session 1: Section 1.
Study session 2: Section 2.
Study session 3: Section 3.
Study session 4: Section 4.
Study session 5: Section 5. You will need access to your computer for this session, together with *Computer Book 1*.
Study session 6: TMA questions on Part I.

Part II

Study session 7: Section 6.
Study session 8: Section 7.
Study session 9: Section 8 and Chapters 3 and 4 of *Computer Book 1*. You will need access to your computer for this session, together with *Computer Book 1*.
Study session 10: Chapter 5 of *Computer Book 1* and TMA questions on Part II. You will need access to your computer for this session, together with *Computer Book 1*.

Part III

Study session 11: Section 10.
Study session 12: Section 11.
Study session 13: Section 12.
Study session 14: Section 13.
Study session 15: TMA questions on Part III and consolidation of your work on this book.

Introduction

People are becoming increasingly aware of the impact of their lifestyle, work and environment on their health. Results of studies suggesting a link between some product or activity and ill health are frequently reported in the media, often raising widespread interest and, occasionally, concern. Some of these reports are controversial, and are disputed, especially when major economic interests are at stake. For example, studies in the 1950s found that smoking was associated with lung cancer; this claim was disputed for many years. Today, after much painstaking research, there is no longer any serious doubt that smoking causes lung cancer.

Many other links have been demonstrated: infection with rubella virus in early pregnancy may cause malformations in the child; populations without access to clean drinking water have higher infant mortality rates; insufficient physical exercise and a diet high in saturated fats are associated with heart disease; wearing a seat belt increases the likelihood of surviving a serious car accident. How such associations are established and interpreted is the topic of this book.

Individuals may respond differently to the same set of circumstances. For example, not everyone who smokes develops lung cancer. However, smoking greatly increases a person's chance of doing so. This can be converted into a statistical statement about probabilities in populations of smokers and non-smokers. The field of medical science that deals with the health of populations is called **epidemiology**, and statistics is central to it. In Parts I and II of this book, you will learn about some of the statistical methods used in epidemiology, and more widely in medical statistics, to investigate associations like those mentioned above.

In recent decades, there has been a big increase in the numbers and types of drugs and treatments available to treat and prevent disease. For example, statins, which have been described as the new aspirin, have been shown to protect against strokes and heart attacks; the death rates from many cancers have been reduced substantially thanks to effective treatments, such as tamoxifen in breast cancer; and effective new treatments are becoming available for rheumatoid arthritis. The evaluation of drugs and treatments raises numerous statistical issues and constitutes an important aspect of modern medical statistics. In Part III, you will learn several statistical methods for evaluating drugs and treatments, and for combining information from several studies. Finally, you will work through an article from the medical literature that uses many of the statistical methods covered in this book.

Part I Cohort studies and case-control studies

Introduction to Part I

The aim of an epidemiological investigation is usually to study the impact on health of one or more potential risk factors. Most often, studies focus on a particular health outcome. This could be a disease, for example colon cancer, or a consequence of disease, such as death from colon cancer, or, more widely, any event with a bearing on health, such as infection or accident. The risk factor is any variable that might be associated with the outcome. This could be age, socio-economic status, exposure to a toxic chemical, diet, and so on. The purpose of the investigation is to determine whether exposure to the potential risk factor is associated with an increase or a decrease in the frequency of the health outcome of interest.

An epidemiological study can help quantify the evidence for or against an association between exposure to a risk factor and occurrence of a disease. However, it cannot generally determine whether an association is causal. The issue of causality is discussed in more detail in Parts II and III, but it is important to be clear from the outset about the limitations of epidemiological investigations.

> Association does not imply causation.

The two most commonly used types of studies in epidemiology are **cohort studies** and **case-control studies**. In Section 1, you will learn about cohort studies, and how to quantify the strength of an association. In Section 2, the binomial model for cohort studies is described, and used to calculate confidence intervals for different measures of association. A different type of study is introduced in Section 3 — the case-control study. In Section 4, the chi-squared test for no association is discussed. Finally, in Section 5, you will use SPSS to apply the methods described in Sections 1 to 4.

1 Cohort studies

In this section, you will learn about cohort studies. In Subsection 1.1, cohort studies are described with several illustrative examples. In Subsection 1.2, two commonly used measures of association are introduced — the relative risk and the odds ratio; their use is discussed in Subsection 1.3.

1.1 What is a cohort study?

Epidemiological investigations aim to study the association, if any, between an exposure to a risk factor and a health outcome or disease. For simplicity, throughout this book we refer to the exposure E and the disease D. The key feature of a cohort study is that it comprises one or several groups of individuals, who are followed over time. The most frequently used type of cohort study is the *controlled cohort study*. This will be introduced using the study described in Example 1.1.

Example 1.1 Hypertension in later life

Some pregnant women are affected by severe gestational hypertension (that is, high blood pressure) known as pre-eclampsia and eclampsia. In a study to investigate the long-term effects of these problems, the proportions of women with hypertension later in life were compared in two groups. One group included 542 women who suffered from gestational hypertension during their first pregnancy. This is the **exposed group**. The other group included 277 women of similar ages to those in the first group, but who did not suffer from these conditions during their first pregnancy. This is the **control group**. ♦

Wilson, B.J., Watson, M.S., Prescott, G.J. et al. (2003) Hypertensive diseases of pregnancy and risk of hypertension and stroke in later life: results from cohort study. *British Medical Journal*, **326**, 845–849.

In Example 1.1, the exposure E is pre-eclampsia or eclampsia during the first pregnancy, and the disease D is hypertension in later life. If experiencing pre-eclampsia or eclampsia during the first pregnancy increases the chance of hypertension later in life then, provided the two groups are similar in other respects, a higher proportion of women would be expected to develop hypertension in the exposed group than in the control group.

This study is a typical example of a **cohort study** of the association, if any, between an exposure E and a disease D. The key features of a cohort study are as follows.

Cohort studies

A **cohort study** to investigate the association between an exposure E and a disease D has the following features.
◇ It includes one group with the exposure E and a comparable control group without exposure E.
◇ The groups are followed over time and the occurrences of disease D in each group are identified.

The word 'cohort' conveys the idea of a group of individuals marching forward. In the context of epidemiology, the group marches forward through time. In Example 1.1 the cohort comprises the $542 + 277 = 819$ women in the study. An important aspect of a cohort study from a statistical perspective is that the exposure status (E or not E) of each individual is treated as fixed, whereas the disease outcome (D or not D) is not known in advance and is regarded as random.

In Roman times a cohort was a military unit, typically comprising 480 soldiers.

This type of cohort study is more precisely a **controlled** cohort study, since it includes a control group comprising individuals who do not have exposure E but are in other respects comparable to the exposed group.

There are many variants on this basic design. For example, a cohort study may include more than one exposed group. Thus, in a study of the effect of environmental pollutants on asthma it might be appropriate to study exposures to several different pollutants. Such studies will be discussed in Subsection 3.3. Similarly, a single cohort may be used to study several outcomes.

Cohort studies without a separate control group are appropriate in some special settings, but will not be considered here.

The data from a cohort study can be arranged in a simple table. The hypertension data for the study described in Example 1.1 are shown in Table 1.1. Note that the table contains not just the numbers, but also concise labelling.

Table 1.1 Gestational pre-eclampsia or eclampsia and hypertension in later life

Exposure category	Hypertension	No hypertension	Total
Pre-eclampsia or eclampsia	327	215	542
No pre-eclampsia or eclampsia	76	201	277

Activity 1.1 Serious self-inflicted injury and compulsory redundancy

A study in New Zealand investigated the relationship between involuntary redundancy and serious self-inflicted injury (such as suicide attempts) leading to hospitalization or death. Two groups of workers were compared. One group comprised 1945 workers made compulsorily redundant from the Whakatu meat-processing plant in the Hawkes Bay region of New Zealand. This plant was closed down in 1986. The other group comprised 1767 workers from the neighbouring Tomoana meat-processing plant that remained open until 1994. The workforces of the two plants were similar in terms of age, sex, ethnicity, and duration of employment. The numbers of workers who were hospitalized or died due to self-inflicted injury between 1986 and 1994 were obtained. There were 14 such cases among the Whakatu workers and 4 among the Tomoana workers.

(a) Identify the exposure E and the 'disease' D in this study.

(b) Identify the exposed group and the control group.

(c) Set out the data from the study in a table similar to Table 1.1, taking care to provide clear labelling in the title, and in the row and column headings.

Keefe, V., Reid, P., Ormsby, C. et al. (2002) Serious health events following involuntary job loss in New Zealand meat processing workers. International Journal of Epidemiology, 31, 1155–1161.

The term 'disease' is used here to mean an adverse state of health.

The cohort studies considered here are *comparative* studies, in that they are designed to provide a comparison between the exposure group and an unexposed control group. If there is no association between exposure E and disease D, then the underlying probabilities of disease D in the exposed group and the control group are the same, and the sample proportions, that is the proportions actually observed, will differ by an amount consistent with chance variation. If, on the other hand, there is a positive association between E and D, then the underlying probability of disease D will be higher in exposed individuals than in unexposed individuals, and the sample proportions are likely to reflect this difference. Similarly, if there is a negative association between E and D, then the underlying probability of disease D will be lower in exposed individuals than in unexposed individuals.

The probability of disease in exposed individuals is denoted $P(D|E)$. The vertical bar indicates that this is a *conditional* probability, in this case, the underlying probability of disease, *given* that the individual is exposed. Similarly, the probability of disease in unexposed individuals is written $P(D|\text{not } E)$, where 'not E' means 'not exposed'.

In general, data from a cohort study may be presented as in Table 1.2. The column corresponding to 'no disease' has been labelled 'not D'.

Table 1.2 A general data table for a cohort study

Exposure category	Disease outcome D	not D	Total
E (exposed group)	a	b	$n_1 = a + b$
not E (control group)	c	d	$n_2 = c + d$

The probabilities of disease in the exposed group and the control group are estimated by the sample proportions:

$$\widehat{P}(D|E) = \frac{a}{n_1}, \quad \widehat{P}(D|\text{not } E) = \frac{c}{n_2}.$$

The hat symbol is used to distinguish the sample estimates from the underlying probabilities.

Example 1.2 Hypertension in later life: results from cohort study

For the hypertension data of Table 1.1, the proportion of the exposed group with hypertension in later life is

$$\widehat{P}(D|E) = \frac{a}{n_1} = \frac{327}{542} \simeq 0.60.$$

This is an estimate of $P(D|E)$, the underlying probability of disease (in this case, hypertension in later life), given exposure (which in this case is pre-eclampsia or eclampsia during the first pregnancy). The proportion of the control group with hypertension in later life is

$$\widehat{P}(D|\text{not } E) = \frac{c}{n_2} = \frac{76}{277} \simeq 0.27.$$

This is an estimate of $P(D|\text{not } E)$, the underlying probability of disease, given no exposure. Since the proportion $\widehat{P}(D|E)$ is greater than the proportion $\widehat{P}(D|\text{not } E)$, this may be evidence that hypertension later in life is associated with pre-eclampsia or eclampsia during the first pregnancy. However, the difference may be due to random variation. Further analysis is required to establish whether there is a link. ♦

Activity 1.2 Seat belts and children's safety in car accidents

In many countries, babies and small children must be secured in specially designed seats when travelling by car. Older children, however, must use the same belts as adults. Concern has been expressed that these belts might injure children owing to the immature anatomy of a child's pelvis. In a study from Canada, data were obtained on 85 children aged between 4 and 14 years sitting in the left-hand back seat (behind the driver) of cars involved in serious accidents. The numbers sustaining at least moderately severe injury among children who were wearing seat belts at the time of the accident and among children who were not wearing seat belts were counted. The data are shown in Table 1.3.

Halman, I., Chipman, M., Parkin, P.C. and Wright, J.G. (2002) Are seat belt restraints as effective in school age children as in adults? A prospective crash study. *British Medical Journal*, **324**, 1123–1125.

Table 1.3 Seat belt use and injury sustained by children aged 4–14

Exposure category	D: sustained at least moderately severe injury Yes	No	Total
Not wearing a seat belt (E)	14	19	33
Wearing a seat belt (not E)	13	39	52

(a) Estimate $P(D|E)$ and $P(D|\text{not } E)$ from Table 1.3, where E is 'not wearing a seat belt'.

(b) Is it correct to conclude from these estimates alone that $P(D|E)$ is greater than $P(D|\text{not } E)$, and hence that not wearing a seat belt is associated with sustaining severe injury in the event of a car crash?

It may not be immediately clear to you that the study in Activity 1.2 is a cohort study, since the population studied consists of children aged 4–14 who were involved in a serious car accident, rather than 'followed over time'. One way to think of it is that the two groups to be compared are specified according to the exposure status of the children (wearing a seat belt or not wearing a seat belt) just before being involved in a serious accident. At this point their exposure status is known, but the outcome is not. The data for this study were assembled after the accidents had occurred, but are analysed so as to recreate the follow-up, like a video being wound back and replayed. This type of study is called a **retrospective** cohort study and is commonly used in epidemiology.

1.2 Measures of association

Intuitively, it is clear that the stronger the association between an exposure E and a disease D, the larger will be the difference between the probabilities of disease with and without the exposure. A measure of association that is commonly used in epidemiology is the **relative risk** RR, which is defined as follows:

$$RR = \frac{P(D|E)}{P(D|\text{not } E)}.$$

In medical statistics, the terms 'risk' and 'probability' are to a large extent interchangeable. Thus, for example, the probability $P(D|E)$ is commonly referred to as the **risk** of disease given exposure. However, note that RR is called the relative *risk* rather than the relative probability. Since the value of RR can be greater than 1, the relative risk RR is *not* a probability.

If there is no association between E and D, then $P(D|E) = P(D|\text{not } E)$, and hence $RR = 1$. Values of RR greater than 1 correspond to **positive** associations, in which presence of the exposure E is associated with an *increase* in the disease risk. Values of RR less than 1 correspond to **negative** associations, in which the presence of the exposure E is associated with a *decrease* in the disease risk.

The relative risk RR is estimated by substituting the sample proportions for the probabilities. Using the notation in Table 1.4, which was introduced in Table 1.2, the estimated relative risk is

$$\widehat{RR} = \frac{\widehat{P}(D|E)}{\widehat{P}(D|\text{not } E)} = \frac{a/n_1}{c/n_2}. \tag{1.1}$$

Table 1.4 A general data table for a cohort study

	D	not D	Total
E	a	b	n_1
not E	c	d	n_2

Example 1.3 Seat belts and relative risk of injury

For the seat belt data of Table 1.3, the estimated relative risk is

$$\widehat{RR} = \frac{\widehat{P}(D|E)}{\widehat{P}(D|\text{not } E)} = \frac{a/n_1}{c/n_2} = \frac{14/33}{13/52} \simeq 1.70.$$

See Activity 1.2.

Thus the sample relative risk is 1.70. This value is greater than 1, suggesting a positive association between not wearing a seat belt and increased risk of at least moderately severe injury. ♦

The interpretation of the relative risk of 1.70 in Example 1.3 is that, in the event of a serious car accident, not wearing a seat belt multiplies the risk of at least moderately severe injury by 1.70. Another way of saying the same thing is that the risk of at least moderately severe injury is increased by 70%.

The relative risk is used frequently in medical statistics, especially when the underlying risks are low. However, it has some drawbacks as a measure of strength of association.

First, the interpretation of the relative risk RR depends not only on the strength of association, but also on the magnitude of the risks involved. To see this, consider, for instance, a disease outcome that occurs in the control group with probability 0.5; that is, $P(D|\text{not } E) = 0.5$. Since the probability of the outcome in the exposed group cannot be greater than 1, that is $P(D|E) \leq 1$, it follows that

$$RR = \frac{P(D|E)}{P(D|\text{not } E)} \leq \frac{1}{0.5} = 2.$$

So, however strongly the exposure is associated with the outcome, the relative risk cannot be greater than 2. In this case, $RR = 2$ must correspond to the strongest possible association. In contrast, if the probability of the disease outcome in the control group were 0.05 (say) then, by a similar argument, $RR \leq 20$, and $RR = 20$ would correspond to the strongest possible association. So, in this case, a relative risk of 2 would certainly not be regarded as the strongest possible association. Thus the interpretation of the relative risk depends on the magnitude of the risks involved.

A second reason why the relative risk is not an ideal measure of strength of association is provided by the results of Activity 1.3.

Activity 1.3 Protective effect of wearing a seat belt

In Activity 1.2 and Example 1.3, the disease D was defined as 'sustained at least moderately severe injury', and the exposure E was defined as 'not wearing a seat belt'. These definitions of D and E are, to some extent, arbitrary choices. Equally reasonably, the disease D^* could be defined as 'avoided moderate or worse injury' in the event of a serious car accident, and the exposure E^* could be defined as 'wearing a seat belt'. Thus D^* and E^* have been obtained from D and E simply by changing the labels of the disease outcomes and the exposure categories: not D is relabelled D^*, and not E is relabelled E^*. In this case, the data would be presented as in Table 1.5.

Table 1.5 Seat belt use and injury avoided by children aged 4–14

Exposure category	D^*: avoided moderate or worse injury		
	Yes	No	Total
Wearing a seat belt (E^*)	39	13	52
Not wearing a seat belt (not E^*)	19	14	33

(a) Use the data in Table 1.5 to estimate the relative risk RR as

$$\widehat{RR} = \frac{\widehat{P}(D^*|E^*)}{\widehat{P}(D^*|\text{not } E^*)}.$$

(b) Compare this estimate with the estimate of 1.70 for the relative risk that was obtained in Example 1.3.

(c) Interpret the relative risk in terms of the chance of avoiding moderate or worse injury. In your view, has the strength of association between seat belt use and moderate or worse injury changed simply by relabelling the disease outcomes and the exposure categories?

Activity 1.3 shows that the relative risk depends on the labelling of the exposure categories and the disease outcomes. If the labels of the exposure categories are switched and the labels of the disease outcomes are switched, then in general the relative risk will change, even though the association it relates to is the same. The only exception to this is when $RR = 1$: in this case, relabelling the disease outcomes and the exposure categories will still produce $RR = 1$.

Ideally, a measure of strength of association should not depend on the way the exposure categories and the disease outcomes are labelled. Such a measure does exist, and is based on the *odds* of an event. For an event A with probability $P(A)$, the **odds of event A** is written $OD(A)$ and is defined by

$$OD(A) = \frac{P(A)}{1 - P(A)}.$$

For example, if you roll a fair die with faces labelled $1, 2, \ldots, 6$, the odds of obtaining a 6, that is the odds of the die coming to rest with the face labelled 6 uppermost, is

$$OD(6) = \frac{P(6)}{1 - P(6)} = \frac{1/6}{1 - 1/6} = 0.2.$$

Odds are also commonly used for betting. In this context, the odds of obtaining a 6 would be reported as '5 to 1 against'.

Note that odds can take any non-negative value. For example, the odds of not obtaining a 6 is

$$OD(\text{not } 6) = \frac{P(\text{not } 6)}{1 - P(\text{not } 6)} = \frac{5/6}{1 - 5/6} = 5.$$

The odds of disease D given exposure E and the odds of D given no exposure E are calculated in exactly the same way:

$$OD(D|E) = \frac{P(D|E)}{1 - P(D|E)} = \frac{P(D|E)}{P(\text{not } D|E)},$$

$$OD(D|\text{not } E) = \frac{P(D|\text{not } E)}{1 - P(D|\text{not } E)} = \frac{P(D|\text{not } E)}{P(\text{not } D|\text{not } E)}.$$

A second measure of strength of association between an exposure E and a disease D is the **odds ratio**, which is denoted OR and defined by

$$OR = \frac{OD(D|E)}{OD(D|\text{not } E)} = \frac{P(D|E) \times P(\text{not } D|\text{not } E)}{P(\text{not } D|E) \times P(D|\text{not } E)}.$$

To calculate the sample odds ratio, note that, in the notation of Table 1.2,

$$\widehat{OD}(D|E) = \frac{\widehat{P}(D|E)}{\widehat{P}(\text{not } D|E)} = \frac{a/n_1}{b/n_1} = \frac{a}{b}$$

and

$$\widehat{OD}(D|\text{not } E) = \frac{\widehat{P}(D|\text{not } E)}{\widehat{P}(\text{not } D|\text{not } E)} = \frac{c/n_2}{d/n_2} = \frac{c}{d}.$$

Hence

$$\widehat{OR} = \frac{\widehat{OD}(D|E)}{\widehat{OD}(D|\text{not } E)} = \frac{a/b}{c/d} = \frac{a \times d}{b \times c}. \tag{1.2}$$

As for the relative risk, if the odds ratio is equal to 1, then there is no association between the exposure E and the disease D. Also as for the relative risk, an odds ratio greater than 1 indicates a positive association, and an odds ratio less than 1 indicates a negative association.

Example 1.4 Odds ratio of injury with and without seat belts

Taking D to denote 'sustained at least moderately severe injury' and E to denote 'not wearing a seat belt', as in Example 1.3, the sample odds are

The data are in Table 1.3.

$$\widehat{OD}(D|E) = \frac{14}{19} \simeq 0.7368$$

and

$$\widehat{OD}(D|\text{not } E) = \frac{13}{39} \simeq 0.3333.$$

So the sample odds ratio is

$$\widehat{OR} = \frac{\widehat{OD}(D|E)}{\widehat{OD}(D|\text{not } E)} \simeq \frac{0.7368}{0.3333} \simeq 2.21.$$

Alternatively, and more directly, using Formula (1.2),

$$\widehat{OR} = \frac{a \times d}{b \times c} = \frac{14 \times 39}{19 \times 13} \simeq 2.21.$$

The estimated odds ratio is greater than 1, indicating a positive association between not wearing a seat belt and sustaining at least moderately severe injury in the event of a serious accident. The corresponding relative risk was 1.70. In general, the odds ratio is further away from 1 than the relative risk. ◆

In Example 1.4 the odds were estimated to four decimal places, whereas in all previous calculations in this section (for example, in Example 1.2) only two decimal places were retained. The reason for this is that the odds in Example 1.4 were then used to obtain the odds ratio. In general, if a quantity is to be used in a later calculation, four decimal places will be retained. Usually, however, final results will be quoted to two decimal places, as for the estimated odds ratio in Example 1.4.

Activity 1.4 Odds ratio for the protective effect of seat belts

Using Table 1.5, calculate the sample odds ratio for avoidance of moderate or worse injury (D^*) for seat belt use (E^*) in children aged 4–14 involved in a serious car accident. Verify that the value is the same as that obtained in Example 1.4.

In Activity 1.3, you saw that, in most instances, relabelling the exposure categories and the disease categories changes the relative risk. Activity 1.4 shows that this is not the case for the odds ratio: switching the labels for both the exposure categories and the disease outcomes does not change the odds ratio. This suggests that of the two measures of strength of association, the odds ratio is the better.

The two measures of association are summarized in the following box.

Measures of association

The relative risk RR and the odds ratio OR are measures of association between an exposure E and a disease D. They are defined by

$$RR = \frac{P(D|E)}{P(D|\text{not } E)}, \quad OR = \frac{P(D|E) \times P(\text{not } D|\text{not } E)}{P(\text{not } D|E) \times P(D|\text{not } E)}.$$

Data from a cohort study may be presented conveniently as in the following table.

Exposure category	Disease outcome		
	D	not D	Total
E (exposed group)	a	b	$n_1 = a + b$
not E (control group)	c	d	$n_2 = c + d$

The relative risk RR and the odds ratio OR may be estimated in a cohort study by

$$\widehat{RR} = \frac{a/n_1}{c/n_2}, \quad \widehat{OR} = \frac{a \times d}{b \times c}.$$

1.3 Which measure of association should be used?

In Subsection 1.2, two measures of association in cohort studies were defined: the relative risk RR and the odds ratio OR. It was argued that the odds ratio represents the better measure of strength of association. This is one reason for preferring OR over RR.

However, measuring the strength of an association in a satisfactory manner is just one aspect of summarizing epidemiological data: communicating the results is also important. The most appropriate measure to use depends on the context. For example, a doctor who needs to convey to her patients the impact of smoking might find it more convenient to quote relative risks than odds ratios. Thus a statement such as 'smoking is associated with an x-fold increase in the risk of lung cancer' is another way of saying that the relative risk of lung cancer in smokers compared to non-smokers is $RR = x$.

For uncommon diseases, the odds ratio and the relative risk are virtually identical. This is because, in the notation of Table 1.4, a is then very much less than n_1, and c is very much less than n_2. Thus $b = n_1 - a \simeq n_1$ and $d = n_2 - c \simeq n_2$, and hence

$$\widehat{OR} = \frac{a \times d}{b \times c} \simeq \frac{a \times n_2}{n_1 \times c} = \frac{a/n_1}{c/n_2} = \widehat{RR}.$$

Activity 1.5 Measures of association for redundancy data

In Activity 1.1 data were presented on serious self-inflicted injury (SSII) in two groups of workers, one made compulsorily redundant and the other not. The data are reproduced in Table 1.6.

Table 1.6 Serious self-inflicted injury (SSII) and compulsory redundancy in meat-processing workers in New Zealand, 1986–94

Exposure category	SSII	No SSII	Total
Made compulsorily redundant (Whakatu workers)	14	1931	1945
Not made compulsorily redundant (Tomoana workers)	4	1763	1767

(a) Use these data to obtain estimates of RR and OR for the association between compulsory redundancy and SSII. What do these indicate about a possible association between compulsory redundancy and SSII?

(b) Comment on the relative sizes of the estimates of RR and OR.

(c) Express the information you obtained in part (a) in one or two sentences, using language appropriate for a non-statistical audience.

Summary of Section 1

Cohort studies for investigating the association between an exposure E and a disease D have been described. A controlled cohort study involves two groups: one group with exposure E and a control group without this exposure. Both groups are followed over time and occurrences of the disease D are recorded. Two measures of association have been introduced: the relative risk RR, and the odds ratio OR. These are used to quantify the strength of association, if any, between an exposure E and a disease D. If there is no association, then $RR = 1$ and $OR = 1$. If there is a positive association between E and D, then $RR > 1$ and $OR > 1$. Similarly, if there is a negative association, then $RR < 1$ and $OR < 1$. If the disease is uncommon, then the relative risk and the odds ratio are similar.

Exercise on Section 1

***Exercise 1.1** Pre-eclampsia or eclampsia and hypertension in later life*

(a) Use the data from Table 1.1 to estimate the relative risk RR and the odds ratio OR for the association between hypertension in later life and pre-eclampsia or eclampsia.

(b) What do these estimates suggest about a possible association between pre-eclampsia or eclampsia, and hypertension later in life?

2 Models for cohort studies

The relative risk RR and the odds ratio OR may be estimated in a cohort study as described in Section 1. However, these estimates are subject to sampling variability: if the cohort study were repeated with different individuals from the same population, the estimates would almost certainly be different. Indeed, if the original study were small, the difference could be large, possibly even suggesting an association in the opposite direction.

***Example 2.1** Post-traumatic stress disorder in Gulf War veterans*

Within months of returning from the 1991 Gulf War, some veterans began to report various symptoms and illnesses. One study from the USA collected data on post-traumatic stress disorder (PTSD) in 6617 veterans who were deployed to the Persian Gulf and 2963 veterans who were deployed to areas other than the Persian Gulf. There were 893 cases of PTSD among veterans deployed to the Gulf, and 180 cases of PTSD among those deployed to other areas. The data are shown in Table 2.1.

Kang, H.K., Natelson, B.H., Mahan, C.M., Lee, K.Y. and Murphy, F.M. (2003) Post-traumatic stress disorder and chronic fatigue syndrome-like illness among Gulf War veterans: a population-based survey of 30 000 veterans. *American Journal of Epidemiology*, **157**, 141–148.

Table 2.1 Deployment to the Gulf and post-traumatic stress disorder (PTSD) in US veterans

Exposure category	PTSD	No PTSD	Total
Deployed to the Gulf	893	5724	6617
Deployed to other areas	180	2783	2963

The estimated odds ratio for association between deployment to the Gulf and PTSD is

$$\widehat{OR} = \frac{893 \times 2783}{5724 \times 180} \simeq 2.41.$$

This is greater than 1, indicating a positive association in this particular sample: among these particular veterans deployment to the Gulf is associated with a higher rate of post-traumatic stress disorder than deployment to other areas. However, to make inferences beyond this particular sample about veterans in general, the uncertainty in the estimate of OR resulting from sampling variability must be quantified. One way to do this is to calculate a confidence interval for OR. ♦

To calculate confidence intervals, a statistical model to represent the random variation in a cohort study must be specified. This is done in Subsection 2.1. Then, in Subsections 2.2 and 2.3, this model is used to derive confidence intervals for the relative risk RR and the odds ratio OR, respectively.

Section 2 Models for cohort studies

2.1 The binomial model

A controlled cohort study involves two groups: a group of individuals with the exposure E, of size n_1, and a control group of individuals without exposure E, of size n_2. These groups are then followed for a specified period and observed for the occurrence of disease D. For individuals in the exposed group the probability of disease during the course of the study is $p_1 = P(D|E)$, and for individuals in the unexposed control group, the probability is $p_2 = P(D|\text{not } E)$.

Let X be a random variable denoting the number of individuals who develop disease D in an exposed group of size n_1, and let Y denote the corresponding number in a control group of size n_2. These variables are displayed in Table 2.2.

Table 2.2   Random variables for a cohort study

Exposure category	Disease outcome		
	D	not D	Total
E (exposed group)	X	$n_1 - X$	n_1
not E (control group)	Y	$n_2 - Y$	n_2

Table 2.2 is similar to Table 1.2, except that the observed frequencies a and c have been replaced by the random variables X and Y.

Provided that the disease outcomes for the individuals in the exposed group and the control group are independent, and that the probability of disease is the same for each individual within each group, then the natural probability models for X and Y are binomial:

$$X \sim B(n_1, p_1), \quad Y \sim B(n_2, p_2).$$

In addition, it will be assumed that X and Y are independent. These assumptions should be checked in each specific application. Activity 2.1 gives an example where some of the assumptions are violated.

Activity 2.1 Air bags and the risk of dying in a car accident

To evaluate the effectiveness of air bags in reducing the probability of dying in a car crash, a study was undertaken based on records of car accidents in the United States. Records were selected of serious accidents involving cars with, in addition to the driver, a single person in the front passenger seat.

There were 8517 cars in which the driver had an air bag and the passenger did not have an air bag. Table 2.3 shows the numbers of fatalities among drivers and passengers.

Cummings, P., McKnight, B., Rivara, F.P. and Grossman, D.C. (2002) Association of driver air bags with driver fatality: a matched cohort study. *British Medical Journal*, **324**, 1119–1122.

Table 2.3 Fatalities among drivers with air bags and passengers without air bags

Exposure category	Died		
	Yes	No	Total
Driver with air bag	4474	4043	8517
Passenger without air bag	5496	3021	8517

(a) Comment on the relation between seat position and having an air bag.

(b) Are the outcome variables X and Y independent? Explain your reasoning.

Comment

In the published analysis, additional data were used to separate out the effects of seat position and air bag use, and the method of analysis allowed for the pairing of drivers and passengers. The conclusion of the study was that air bags reduced the risk of death by about 8%, compared to a reduction of 65% associated with seat belts. Using both reduced the risk of death by 68%.

2.2 Confidence intervals for the relative risk

In this subsection approximate confidence intervals for the relative risk are obtained. Since, in practice, it is simpler to work with the logarithm of the relative risk than with the relative risk itself, the method used involves first finding a z-interval for the logarithm of the relative risk, and then calculating the corresponding confidence interval for the relative risk.

The notation introduced in Table 1.2 for the entries in a data table for a cohort study will be used throughout this subsection. It is reproduced in Table 2.4 for ease of reference.

Table 2.4 A general data table for a cohort study

	D	not D	Total
E	a	b	n_1
not E	c	d	n_2

In general, for a sufficiently large sample, an approximate $100(1-\alpha)\%$ confidence interval (or z-interval) for a parameter θ, which is denoted (θ^-, θ^+), is given by

$$(\theta^-, \theta^+) = (\widehat{\theta} - z\widehat{\sigma}, \widehat{\theta} + z\widehat{\sigma}),$$

where $\widehat{\theta}$ is the sample estimate of θ, $\widehat{\sigma}$ is the estimated standard error of the estimator $\widehat{\theta}$ and z is the $(1-\alpha/2)$-quantile of the standard normal distribution.

Let $\theta = \log(RR)$. The estimate of the relative risk is given by Formula (1.1):

$$\widehat{RR} = \frac{a/n_1}{c/n_2}.$$

So the estimate $\widehat{\theta}$ of θ is $\log(\widehat{RR})$.

If σ denotes the standard error of the estimator $\widehat{\theta} = \log(\widehat{RR})$ then, for n_1 and n_2 sufficiently large, it can be shown that σ can be estimated by

$$\widehat{\sigma} = \sqrt{\frac{1}{a} - \frac{1}{n_1} + \frac{1}{c} - \frac{1}{n_2}}. \tag{2.1}$$

The proof uses a mathematical technique known as Taylor series expansion and, while it is not difficult, it will be omitted.

Thus an approximate $100(1-\alpha)\%$ confidence interval for $\theta = \log(RR)$ is given by

$$(\theta^-, \theta^+) = \left(\log(\widehat{RR}) - z \times \widehat{\sigma}, \log(\widehat{RR}) + z \times \widehat{\sigma}\right),$$

where $\widehat{\sigma}$ is given by Formula (2.1) and z is the $(1-\alpha/2)$-quantile of the standard normal distribution.

Since $\theta = \log(RR)$, it follows that $RR = \exp(\theta)$, and hence an approximate $100(1-\alpha)\%$ confidence interval (RR^-, RR^+) for the relative risk RR is $(\exp(\theta^-), \exp(\theta^+))$. Thus

$$RR^- = \exp\left(\log(\widehat{RR}) - z \times \widehat{\sigma}\right) = \widehat{RR} \times \exp(-z \times \widehat{\sigma}),$$

$$RR^+ = \exp\left(\log(\widehat{RR}) + z \times \widehat{\sigma}\right) = \widehat{RR} \times \exp(z \times \widehat{\sigma}).$$

So an approximate $100(1-\alpha)\%$ confidence interval for the relative risk RR is given by

$$(RR^-, RR^+) = \left(\widehat{RR} \times \exp(-z \times \widehat{\sigma}), \widehat{RR} \times \exp(z \times \widehat{\sigma})\right),$$

where $\widehat{\sigma}$ is given by (2.1) and z is the $(1-\alpha/2)$-quantile of $N(0,1)$.

The details are summarized in the following box.

Confidence intervals for the relative risk

The sample estimate $\widehat{RR}$ of the relative risk RR derived from a cohort study is, using the notation of Table 2.4, given by

$$\widehat{RR} = \frac{a/n_1}{c/n_2}. \qquad (2.2)$$

For sufficiently large n_1 and n_2, an approximate $100(1-\alpha)\%$ confidence interval for the relative risk RR is

$$(RR^-, RR^+) = \left(\widehat{RR} \times \exp(-z \times \hat{\sigma}), \widehat{RR} \times \exp(z \times \hat{\sigma})\right), \qquad (2.3)$$

where z is the $(1-\alpha/2)$-quantile of the standard normal distribution and

$$\hat{\sigma} = \sqrt{\frac{1}{a} - \frac{1}{n_1} + \frac{1}{c} - \frac{1}{n_2}}. \qquad (2.4)$$

Example 2.2 Breast cancer and hormone-replacement therapy

Concerns over a possible link between breast cancer and hormone-replacement therapy (HRT) led to a very large cohort study being undertaken in the UK. Between 1996 and 2001, the study, known as the Million Women Study, recruited 1 084 110 women aged between 50 and 64 who were followed up for cancer. Table 2.5 contains data from this study on two groups of women: women who were using combined oestrogen-progestagen HRT at the time of recruitment (the exposed group), and women who had never used HRT at the time of recruitment (the control group). In each group, the number of new cases of invasive breast cancer occurring during the study follow-up period was counted.

Million Women Study Collaborators (2003) Breast cancer and hormone-replacement therapy in the Million Women Study. *Lancet*, **362**, 419–427.

Table 2.5 Invasive breast cancer and use of oestrogen-progestagen HRT

Exposure category	Invasive breast cancer		Total
	Yes	No	
Currently using combined oestrogen-progestagen HRT	1934	140 936	142 870
Never used HRT	2894	389 863	392 757

The estimated relative risk of invasive breast cancer for use of oestrogen-progestagen HRT is

$$\widehat{RR} = \frac{a/n_1}{c/n_2} = \frac{1934/142\,870}{2894/392\,757} \simeq 1.8371.$$

For a 99% confidence interval, the 0.995-quantile of the standard normal distribution is required. From the table of quantiles of the standard normal distribution in the *Handbook*, this is $z = 2.576$.

The estimated standard error $\hat{\sigma}$ is given by (2.4):

$$\hat{\sigma} = \sqrt{\frac{1}{a} - \frac{1}{n_1} + \frac{1}{c} - \frac{1}{n_2}}$$

$$= \sqrt{\frac{1}{1934} - \frac{1}{142\,870} + \frac{1}{2894} - \frac{1}{392\,757}}$$

$$\simeq 0.02921.$$

So, using (2.3), the 99% confidence limits for the relative risk are

$$RR^- = \widehat{RR} \times \exp(-z \times \hat{\sigma})$$

$$\simeq 1.8371 \times \exp(-2.576 \times 0.02921)$$

$$\simeq 1.70,$$

$$RR^+ = \widehat{RR} \times \exp(z \times \hat{\sigma})$$

$$\simeq 1.8371 \times \exp(2.576 \times 0.02921)$$

$$\simeq 1.98.$$

Thus the risk of invasive breast cancer is 1.84 times higher in women taking oestrogen-progestagen HRT than in women who never took any HRT, with 99% confidence interval $(1.70, 1.98)$. The confidence interval is quite narrow, reflecting the large size of the study, and is located entirely above 1. This indicates a positive association between HRT use and breast cancer. ♦

Activity 2.2 Efficacy of measles vaccines

Prior to the introduction of routine childhood vaccination against measles in 1968, there were hundreds of thousands of cases of measles every year in the UK. In 1964, a cohort study was undertaken to evaluate the efficacy of measles vaccines. Children aged between 10 months and 2 years were enrolled into one of three groups: an unvaccinated group who received no vaccine, a vaccinated group who received live measles vaccine, and a second vaccinated group who received killed measles vaccine followed by live measles vaccine. Allocation to the groups was based on day of birth. Table 2.6 shows the numbers of children in the unvaccinated and live vaccine groups (the live vaccine was the one chosen subsequently for routine immunization), and the numbers of measles cases arising within six months after vaccination.

Medical Research Council (1966) Vaccination against measles: a clinical trial of live measles vaccine given alone and live vaccine preceded by killed vaccine. *British Medical Journal*, 19 February, 441–446.

Table 2.6   Measles vaccination and measles infection

Exposure category	Measles within six months		
	Yes	No	Total
Received live measles vaccine	156	9 421	9 577
Unvaccinated	1531	14 797	16 328

(a) Estimate the relative risk of measles after vaccination.

(b) Obtain a 99% confidence interval for the relative risk RR.

(c) Interpret your results. Does the live vaccine protect against measles?

2.3 Confidence intervals for the odds ratio

In this subsection, approximate confidence intervals for the odds ratio are obtained. The method for calculating confidence intervals for the odds ratio is very similar to that for the relative risk. The derivation of the formula for a confidence interval also involves using logarithms, and the formula for an approximate confidence interval for the odds ratio is similar in form to that for the relative risk. The main difference is in the formula for the estimated standard error $\hat{\sigma}$ that is required to calculate a confidence interval. And, of course, the formula for the estimate $\widehat{OR}$ is different from that for $\widehat{RR}$. The details are given in the following box.

Confidence intervals for the odds ratio

The sample estimate $\widehat{OR}$ of the odds ratio derived from a cohort study is, using the notation of Table 2.4, given by

$$\widehat{OR} = \frac{a \times d}{b \times c}. \tag{2.5}$$

For sufficiently large $n_1 = a + b$ and $n_2 = c + d$, an approximate $100(1 - \alpha)\%$ confidence interval for the odds ratio OR is

$$(OR^-, OR^+) = \left(\widehat{OR} \times \exp(-z \times \hat{\sigma}), \widehat{OR} \times \exp(z \times \hat{\sigma})\right), \tag{2.6}$$

where z is the $(1 - \alpha/2)$-quantile of the standard normal distribution and

$$\hat{\sigma} = \sqrt{\frac{1}{a} + \frac{1}{b} + \frac{1}{c} + \frac{1}{d}}. \tag{2.7}$$

Example 2.3 *Cannabis use and mental health*

Recreational use of cannabis is now widespread among young people in many countries. Uncertainty persists about its consequences for health. Some research has suggested that heavy use of cannabis is associated with depression and anxiety. A cohort study was undertaken in Australia to investigate this hypothesis. Teenagers were recruited from 44 schools in Victoria, Australia, and followed to age 20 or 21. Cannabis use was monitored using self-administered questionnaires.

Patton, G.C., Coffey, C., Carlin, J.B., Degenhardt, L., Lynskey, M. and Hall, W. (2002) Cannabis use and mental health in young people: cohort study. *British Medical Journal*, **325**, 1195–1198.

Table 2.7 shows data for young women derived from this study. Frequency of cannabis use was grouped into two categories: less than weekly (the control group), and weekly or more (the exposed group). The outcome of interest (D) is depression or anxiety, assessed at interviews with specialists.

Table 2.7 Depression and anxiety in young women according to cannabis use

Frequency of cannabis use	Depression or anxiety Yes	No	Total
Weekly or more	35	34	69
Less than weekly	153	637	790

The estimated odds ratio of depression or anxiety for frequent use of cannabis compared to infrequent use among young women is given by (2.5):

$$\widehat{OR} = \frac{a \times d}{b \times c} = \frac{35 \times 637}{34 \times 153} \simeq 4.2859.$$

For a 95% confidence interval, the 0.975-quantile of the standard normal distribution is required; this is $z = 1.96$.

The estimated standard error $\hat{\sigma}$ is given by (2.7):

$$\hat{\sigma} = \sqrt{\frac{1}{a} + \frac{1}{b} + \frac{1}{c} + \frac{1}{d}}$$

$$= \sqrt{\frac{1}{35} + \frac{1}{34} + \frac{1}{153} + \frac{1}{637}}$$

$$\simeq 0.2571.$$

So, using (2.6), the 95% confidence limits for the odds ratio are

$$OR^- = \widehat{OR} \times \exp(-z \times \hat{\sigma})$$

$$\simeq 4.2859 \times \exp(-1.96 \times 0.2571)$$

$$\simeq 2.59,$$

$$OR^+ = \widehat{OR} \times \exp(z \times \hat{\sigma})$$

$$\simeq 4.2859 \times \exp(1.96 \times 0.2571)$$

$$\simeq 7.09.$$

Thus the odds ratio is 4.29, with 95% confidence interval (2.59, 7.09). The confidence interval for OR is located well above 1, suggesting that heavy cannabis use is associated with a big increase in the odds of depression or anxiety in young women. ♦

Activity 2.3 Cannabis use and depression or anxiety in young men

The data in Example 2.3 relate only to young women. The data on young men from the same study are shown in Table 2.8.

Table 2.8 Depression and anxiety in young men according to cannabis use

Frequency of cannabis use	Depression or anxiety Yes	No	Total
Weekly or more	20	126	146
Less than weekly	51	534	585

(a) Estimate the odds ratio of anxiety or depression for frequent use of cannabis relative to infrequent use among young men.

(b) Obtain a 95% confidence interval for the odds ratio.

(c) Interpret your findings.

Summary of Section 2

The binomial model for cohort studies has been introduced. The assumptions of this model are that outcomes for individuals within each group are independent and occur with the same probability. To calculate confidence intervals, it is also assumed that outcomes are independent between groups. Approximate confidence intervals for RR and OR have been presented, based on the logarithm of the relative risk and odds ratio.

Exercise on Section 2

Exercise 2.1 Post-traumatic stress disorder

(a) In Example 2.1 data were presented on post-traumatic stress disorder (PTSD) in US veterans and the odds ratio OR was estimated. Use the data in Table 2.1 to obtain a 99% confidence interval for OR.

(b) Is it plausible that deployment to the Gulf is not associated with increased rates of PTSD in US veterans? Explain carefully why you reach your conclusion.

3 Case-control studies

One drawback of cohort studies is that, when the disease of interest is uncommon, a cohort study may need to be very large or involve very lengthy follow-up in order to obtain sufficient numbers of individuals with the disease. For example, if a health event occurs on average in one per thousand individuals over a given period, then to obtain ten cases would require about 10 000 individuals to be followed over that period. Such large studies are usually time-consuming and costly to undertake.

In this section, a different study design, the **case-control** study, is considered. This study design makes it possible to study uncommon health outcomes without the need for very large samples or very lengthy studies. In Subsection 3.1, case-control studies are described, and in Subsection 3.2, measures of association for case-control studies are discussed. In Subsection 3.3, studies with several exposure categories are considered.

3.1 What is a case-control study?

You have seen that, in a typical controlled cohort study, two groups of individuals — an exposed group with exposure E and a control group without exposure E — are followed over time and occurrences of disease D are counted within each group. The data from such a study may be summarized as in Table 3.1.

Table 3.1 A general data table for a cohort study

Exposure category	Disease outcome		Total
	D	not D	
E (exposed group)	a	b	$n_1 = a + b$
not E (control group)	c	d	$n_2 = c + d$

A key aspect of the cohort study design is the difference between the status of the exposure E and the disease D. For each individual in the study, the exposure category is regarded as *fixed* and the disease outcome is regarded as *random*.

In a case-control study, a sample of cases — that is, individuals who have the disease D (for example, women with breast cancer) — is selected. Then a second group of individuals who are not cases (for example, women who do not have breast cancer) is selected. These individuals are the controls. Thus in these two groups, cases and controls, the disease outcome is known. However, the exposure category of the cases and the controls is treated as random. After the cases and controls have been selected, their previous exposures (for example, whether they ever used hormone-replacement therapy) are ascertained. This results in a data table as set out in Table 3.2.

Table 3.2 A general data table for a case-control study

Exposure category	Disease outcome	
	D (cases)	not D (controls)
E	a	b
not E	c	d
Total	$m_1 = a + c$	$m_2 = b + d$

Note the differences between Table 3.2 and Table 3.1. In a case-control study (Table 3.2), the numbers with disease D and without disease D are fixed in advance: the study includes m_1 cases and m_2 controls. In contrast, in a cohort study, the exposure group and the control group, of sizes n_1 and n_2, are fixed in advance. In a case-control study (Table 3.2), the presence or absence of exposure E is determined by looking back in time at the histories of the cases and controls. In contrast, in a cohort study the disease outcome is determined by following the exposed group and the unexposed group forward in time. The key features of a case-control study are set out in the following box.

Case-control studies

A case-control study of the association between an exposure E and a disease D has the following features.

◇ It includes a group of cases with the disease D and a group of controls without the disease D who are otherwise comparable to the cases.

◇ The past exposures of cases and controls are determined and occurrences of exposure E are identified.

An important issue in case-control studies is how to select the controls. As a general rule, they should be selected from the population that gave rise to the cases, and should have had the same opportunity as the cases to become exposed.

Once exposures in cases and controls have been determined, the proportions of cases and controls with exposure E are compared. If there is no association between E and D, then the proportions exposed should be similar in cases and controls. On the other hand, if D and E are positively associated, a greater proportion of cases than controls would be expected to have exposure E.

Example 3.1 *Smoking and lung cancer*

The 1950 case-control study of smoking and lung cancer by Richard Doll and Austin Bradford Hill is a classic example. An unexplained increase in lung cancer deaths had taken place over the previous decades in the UK and several other countries. The increase was spectacular: lung cancer deaths in the UK had increased fifteen-fold between 1922 and 1947. Several exposures had been suggested as possible causes for the increase, including industrial pollution, tarred roads, exhaust fumes from cars, as well as smoking.

Doll, R. and Hill, A.B. (1950) Smoking and carcinoma of the lung: preliminary report. *British Medical Journal*, 30 September, 739–748.

Doll and Hill investigated the association by means of a case-control study. The cases were patients admitted to hospital with lung cancer. For each case, a control of the same sex and similar age admitted to the hospital for a disease other than cancer was selected.

Smoking histories were then obtained for each of the 709 cases and 709 controls. Table 3.3 shows the data for the males (649 cases and 649 controls), with exposure defined as having been a smoker at any time in the past.

Table 3.3   Smoking and lung cancer in males

Exposure category	Cases of lung cancer	Controls
Smoked	647	622
Never smoked	2	27
Total	649	649

An important feature of the case-control design is that it involves only 649 cases and 649 controls. A cohort study would have had to be very large to result in 649 lung cancer cases.

The proportion of smokers among cases is $647/649 \simeq 0.9969$ compared to $622/649 \simeq 0.9584$ among controls. Thus it appears that smoking was extremely widespread, but more common among lung cancer cases than among controls. In Subsection 3.2, a suitable measure of association will be discussed. ♦

In Example 3.1 the numbers of controls and cases were the same. This is not a requirement: the numbers of cases and controls are often different, as in the study described in Activity 3.1.

Activity 3.1   Political activity and homicide in Karachi

In 2001 it was estimated that every year half a million people are murdered in the world. In Karachi, Pakistan, homicide rates vary substantially between neighbourhoods. A case-control study was undertaken in one area to identify whether political activity was associated with death by homicide.

Mian, A., Mahmood, S.F., Chotani, H. and Luby, S. (2002) Vulnerability to homicide in Karachi: political activity as a risk factor. *International Journal of Epidemiology*, **31**, 581–585.

Altogether 35 victims of homicide were included in the study, and 85 controls with similar age and sex distribution as the victims. Household members were questioned about the political activities of the study subjects. Of the 35 victims, eleven had attended political meetings, compared to two of the controls.

(a) Arrange these data in a table similar to Table 3.3, indicating clearly what the exposure is.

(b) Informally compare the proportions of exposed cases (victims of homicide) and exposed controls. What does this suggest?

3.2   Measures of association in case-control studies

In a cohort study, the numbers exposed and not exposed are regarded as fixed, and the measures of association, namely the relative risk RR and the odds ratio OR, are based on $P(D|E)$, the probability of disease given exposure, and $P(D|\text{not } E)$, the probability of disease given no exposure.

In a case-control study, however, the numbers of cases (with disease D) and controls (without disease D) included are decided in advance by the investigator. In consequence, $P(D|E)$ and $P(D|\text{not } E)$ cannot be estimated in a case-control study. So the relative risk RR, which is the ratio of these probabilities, cannot be estimated in a case-control study. However, the odds ratio can be estimated. This is illustrated in Example 3.2.

Example 3.2 Measures of association in case-control studies

In Example 3.1, data from the famous Doll and Hill case-control study of smoking and lung cancer were presented. These data are reproduced in Table 3.4. Also included in the table are the row totals that would be presented if this were a cohort study.

Table 3.4 Smoking and lung cancer in males

Exposure category	Cases of lung cancer	Controls	Total
Smoked	647	622	1269
Never smoked	2	27	29
Total	649	649	1298

Suppose that the relative risk and odds ratio were to be estimated as if this were a cohort study. Then the estimates would be as follows:

$$\widehat{RR} = \frac{647/1269}{2/29} \simeq 7.39, \quad \widehat{OR} = \frac{647 \times 27}{622 \times 2} \simeq 14.04.$$

Now suppose that Doll and Hill had, in fact, decided to obtain ten times as many controls as cases. This is perfectly admissible: the numbers of cases and controls chosen is entirely within the control of the investigators. Then, assuming that the proportion of smokers among these controls was the same as among those actually selected, the data would have been as below.

Exposure category	Cases of lung cancer	Controls ×10	Total
Smoked	647	6220	6867
Never smoked	2	270	272
Total	649	6490	7139

This would produce the following estimates of the relative risk and odds ratio:

$$\widehat{RR} = \frac{647/6867}{2/272} \simeq 12.81, \quad \widehat{OR} = \frac{647 \times 270}{6220 \times 2} \simeq 14.04.$$

The relative risk has increased, but the odds ratio is unchanged. Similarly, if Doll and Hill had used ten times as many cases as they did, the data would have been as follows.

Exposure category	Cases of lung cancer ×10	Controls	Total
Smoked	6470	622	7092
Never smoked	20	27	47
Total	6490	649	7139

The estimated relative risk and odds ratio in this case would be as follows:

$$\widehat{RR} = \frac{6470/7092}{20/47} \simeq 2.14, \quad \widehat{OR} = \frac{6470 \times 27}{622 \times 20} \simeq 14.04.$$

The relative risk is now much lower, but again the odds ratio is unchanged at 14.04. In fact, the odds ratio will be the same whatever the ratio of cases to controls, whereas the relative risk will vary. ♦

Example 3.2 shows that the relative risk RR *cannot* be estimated in a case-control study, since its value varies according to how many cases and controls are selected. However, the odds ratio OR does not depend on how many cases and controls are selected, only on the proportions of cases and controls with the exposure. In particular, its estimate from a case-control study would have the same value as if a cohort study had been undertaken. Thus the odds ratio *can* be estimated in a case-control study. This may be demonstrated more generally as follows.

Section 3 Case-control studies

In a case-control study it is possible to estimate $OD(E|D)$, the odds of exposure in cases, and $OD(E|\text{not } D)$, the odds of exposure in controls. The estimates, using the notation of Table 3.2, are as follows:

$$\widehat{OD}(E|D) = \frac{a/m_1}{c/m_1} = \frac{a}{c},$$

$$\widehat{OD}(E|\text{not } D) = \frac{b/m_2}{d/m_2} = \frac{b}{d}.$$

Clearly, it is also possible to estimate the ratio of these odds:

$$\frac{\widehat{OD}(E|D)}{\widehat{OD}(E|\text{not } D)} = \frac{a/c}{b/d} = \frac{a \times d}{b \times c}.$$

Notice that this expression is the same as that obtained for the estimate of the odds ratio OR in a cohort study (see Result (1.2)). It follows that

$$\widehat{OR} = \frac{\widehat{OD}(D|E)}{\widehat{OD}(D|\text{not } E)} = \frac{\widehat{OD}(E|D)}{\widehat{OD}(E|\text{not } D)}.$$

This identity also applies to the underlying parameters, that is, without the 'hat' symbols.

Hence the odds ratio OR can be estimated in a case-control study, even though the relative risk RR cannot. Confidence intervals for OR are also calculated in the same way in a case-control study as in a cohort study. These facts are summarized in the following box.

Measures of association in case-control studies

In a case-control study, the odds ratio OR can be estimated but the relative risk RR cannot. Using the notation of Table 3.2, the odds ratio is estimated by

$$\widehat{OR} = \frac{a \times d}{b \times c}.$$

Approximate $100(1 - \alpha)\%$ confidence intervals for OR are calculated in the same way in a case-control study as in a cohort study (see Formulas (2.6) and (2.7)).

In Subsection 1.3 you saw that, for uncommon diseases, the odds ratio and the relative risk are in fact very close. Thus, in case-control studies, although the relative risk cannot be estimated directly, when the disease is uncommon, it can be approximated by the odds ratio.

Example 3.3 Alcohol consumption and fatal car accidents

The twentieth century saw the emergence of a new and deadly epidemic: injury from car accidents. Alcohol consumption was soon identified as a likely cause of accidents. This example is based on the first controlled study of the role of alcohol consumption in causing fatal car accidents. The study design chosen was the case-control design.

McCarroll, J.R. and Haddon, W. (1962) A controlled study of fatal automobile accidents in New York City. Journal of Chronic Diseases, 15, 811–826.

Details were obtained of all fatalities from car accidents in New York between June and October in 1959 and in 1960. The fatalities were classified according to whether or not the dead person was considered to be responsible for the accident. This example includes 24 drivers who were killed in car accidents for which they were considered to be responsible. This group of 24 constitutes the cases. Their blood alcohol levels were obtained from post-mortem examinations.

Controls were obtained by selecting drivers passing the locations where the accidents of the cases occurred, at the same time of day and on the same day of the week. A total of 154 controls were selected in this way. The controls were breathalyzed. Exposure is defined as a high blood alcohol level, namely a concentration greater than or equal to 100 mg% (1 mg% is 1 mg of alcohol per 100 ml of blood). The data are in Table 3.5 (overleaf).

The study authors, who were accompanied by police, report encountering 'occasional hostility, and in one case an initial plea of diplomatic immunity'.

The estimated odds ratio for the association between alcohol level and dying in a car accident for which one is responsible is given by

$$\widehat{OR} = \frac{a \times d}{b \times c} = \frac{14 \times 146}{8 \times 10} = 25.55.$$

Thus the odds of causing a car accident and dying in it are 25.55 times higher for drivers with high blood alcohol levels than for other drivers.

Table 3.5 Alcohol levels and fatal car accidents

Alcohol level	Cases	Controls
$\geq 100\,\text{mg}\%$	14	8
$< 100\,\text{mg}\%$	10	146
Total	24	154

For a 95% confidence interval, the 0.975-quantile of the standard normal distribution is required: $z = 1.96$. The estimated standard error $\hat{\sigma}$ is given by (2.7):

$$\hat{\sigma} = \sqrt{\frac{1}{a} + \frac{1}{b} + \frac{1}{c} + \frac{1}{d}}$$

$$= \sqrt{\frac{1}{14} + \frac{1}{8} + \frac{1}{10} + \frac{1}{146}}$$

$$\simeq 0.5507.$$

So the confidence limits are

$$OR^- = \widehat{OR} \times \exp(-z \times \hat{\sigma})$$
$$\simeq 25.55 \times \exp(-1.96 \times 0.5507)$$
$$\simeq 8.68,$$

$$OR^+ = \widehat{OR} \times \exp(z \times \hat{\sigma})$$
$$\simeq 25.55 \times \exp(1.96 \times 0.5507)$$
$$\simeq 75.19.$$

The confidence interval for the odds ratio is $(8.68, 75.19)$, indicating a positive (and rather strong) association between alcohol consumption and fatal car accidents. ♦

Activity 3.2 Cot deaths and sleeping position

Sudden unexplained deaths in apparently normal babies under one year of age are known as sudden infant deaths, or cot deaths. In the UK, they are the leading cause of death in babies aged between one month and one year. The causes of Sudden Infant Death Syndrome (SIDS) are not known. In 1990 a case-control study was published that suggested that babies who were put down to sleep on their front and who were too heavily wrapped were more likely to die of SIDS. Following this study, the 'Back to Sleep' campaign was launched in several countries to encourage parents to place their babies to sleep on their back and to avoid overheating and smoky environments. In subsequent years, deaths from SIDS dropped by over 50%. Table 3.6 shows data from the 1990 study.

Fleming, P.J., Gilbert, R., Azaz, Y. et al. (1990) Interaction between bedding and sleeping position in the sudden infant death syndrome: a population based case-control study. *British Medical Journal*, **301**, 85–89.

Table 3.6 Sleeping position and deaths from SIDS

Position baby last placed down to sleep	Cases	Controls
On its front	62	76
In another position	5	55
Total	67	131

A total of 67 babies who died of SIDS were included. The controls are live babies of similar ages and from the same localities as the babies that died. The exposure is last placing the baby down to sleep on its front.

(a) Estimate the odds ratio of SIDS associated with the front sleeping position.

(b) Calculate a 95% confidence interval for the odds ratio.

(c) Interpret your results.

3.3 Studies with more than two exposure categories

So far, all the studies we have considered (whether cohort or case-control) have involved just two exposure categories: exposed and unexposed. In some cases, however, there may be more than two exposure categories. In this subsection, you will learn how to analyse data from such studies. The methods described apply to cohort studies as well as case-control studies.

For studies involving more than two exposure categories, it is common practice to identify one category as a **reference** exposure category, and calculate odds ratios for the other exposure categories relative to this reference category. The reference category is usually chosen to represent lack of exposure, or 'normal' exposure in some sense, though the choice is to some extent arbitrary. The choice of reference category depends on the context of each study. The procedure is illustrated in Example 3.4.

Example 3.4 Sleeping position and SIDS

After the success of the 'Back to Sleep' campaign (see Activity 3.2), further research was undertaken to examine the relationship between sleeping position and sudden infant death syndrome (SIDS). Between 1993 and 1995, a case-control study, similar to that described in Activity 3.2, was undertaken with the aim of investigating further the causes of SIDS.

Data for 188 cases and 774 controls were obtained on the position the baby was placed down to sleep: back, side or front. Thus there are three exposure categories, rather than two. The data are in Table 3.7.

Fleming, P.J., Blair, P.S., Bacon, C. *et al.* (1996) Environment of infants during sleep and risk of the sudden infant death syndrome: results of 1993–5 case-control study for confidential inquiry into stillbirths and deaths in infancy. *British Medical Journal*, **313**, 191–195.

Table 3.7 Sleeping position and deaths from SIDS

Position baby last placed down to sleep	Cases	Controls
On its front	30	24
On its side	76	241
On its back	82	509
Total	188	774

To investigate the association between sleeping position and SIDS in a table such as this, one of the exposure categories is chosen as the reference category. In this example, the back position will be chosen (rather arbitrarily) as reference.

Odds ratios for the other two sleeping positions are then calculated, relative to the reference category. For example, for the front position, the odds ratio is calculated using rows 1 and 3 of Table 3.7. Thus

$$\widehat{OR}_{\text{front}} = \frac{30 \times 509}{24 \times 82} \simeq 7.7591.$$

A 95% confidence interval for this odds ratio is calculated in the same way as for a 2×2 table, ignoring the data in the row corresponding to the side sleeping position. Thus the estimated standard error $\widehat{\sigma}$ is given by

$$\widehat{\sigma} = \sqrt{\frac{1}{30} + \frac{1}{24} + \frac{1}{82} + \frac{1}{509}} \simeq 0.2986.$$

It follows that the 95% confidence interval for OR_{front} is $(4.32, 13.93)$. This confirms the findings described in Activity 3.2: the front sleeping position is associated with higher rates of SIDS.

This confidence interval was calculated using (2.6).

The calculation is repeated for the side sleeping position, again relative to the reference exposure category. In this case, the data in the second and third rows of Table 3.7 are used. Thus

$$\widehat{OR}_{\text{side}} = \frac{76 \times 509}{241 \times 82} \simeq 1.9575, \quad \hat{\sigma} = \sqrt{\frac{1}{76} + \frac{1}{241} + \frac{1}{82} + \frac{1}{509}} \simeq 0.1774.$$

The 95% confidence interval for OR_{side} is $(1.38, 2.77)$. Thus the side sleeping position is also associated with an increased risk of SIDS, though the association is not as strong as for the front position. ◆

Activity 3.3 Ectopic pregnancy and genital infection

A pregnancy is called ectopic if the baby develops outside the womb. An ectopic pregnancy is a life-threatening condition requiring emergency treatment. A large case-control study was undertaken in the Auvergne, France, to study the factors associated with ectopic pregnancy.

Bouyer, J., Coste, J., Shojaei, T. *et al.* (2003) Risk factors for ectopic pregnancy: a comprehensive analysis based on a large case-control, population-based study in France. *American Journal of Epidemiology*, **157**, 185–194.

The data in Table 3.8 include 780 women who experienced an ectopic pregnancy, and 1673 control women who gave birth normally. The exposure considered here is history of genital infections. These exposures are classified in three categories: PID (standing for Pelvic Inflammatory Disease, a particular type of genital infection), Non-PID (genital infections other than PID), and None (no history of genital infections).

Table 3.8 Ectopic pregnancy and history of genital infections

History of genital infections	Cases	Controls
PID	212	112
Non-PID	157	407
None	411	1154
Total	780	1673

(a) Identify a suitable reference exposure category.

(b) Estimate the odds ratios and 95% confidence intervals for the other two exposure categories, relative to this reference category.

(c) Interpret your findings.

The procedure for larger tables described in this subsection applies also to cohort studies. For cohort studies, relative risks can be estimated relative to a reference category as well as odds ratios.

Summary of Section 3

Case-control studies have been described and contrasted with cohort studies. In a case-control study, a group of cases with the disease D is compared with a group of controls without the disease. The previous exposures E for cases and controls are then documented. For uncommon diseases, cohort studies may need to be very large and in such circumstances a case-control study may be more practical. You have seen that the relative risk RR cannot be estimated in a case-control study, but the odds ratio OR can. The odds ratio is estimated and confidence intervals are calculated in the same way as for cohort studies. If the disease is uncommon, the relative risk can be approximated by the odds ratio. Studies involving more than two exposure categories are analysed by identifying a reference exposure category and calculating odds ratios relative to this reference category.

Exercise on Section 3

Exercise 3.1 Hib meningitis

Haemophilus Influenzae type b (Hib) causes meningitis. Hib meningitis is a rare but serious disease, occurring most frequently in young children. In the UK, most but not all children are vaccinated against Hib in the first few months of life. In 2003, concerns were expressed about the efficacy of the vaccine used in the UK.

(a) Describe briefly the design of a cohort study to investigate the association between receipt of Hib vaccine (the exposure E) and subsequent Hib meningitis (the disease D).

(b) Describe briefly the design of a case-control study to investigate this association.

(c) State one advantage of the case-control method compared to the cohort method.

(d) An estimate of the relative risk RR is required. How might this be approximated in a case-control study?

4 Testing for no association in cohort studies and case-control studies

In Sections 1 to 3, you have seen that an estimate of a suitable measure of association and a confidence interval for the measure can be used to summarize the strength of association between an exposure E and a disease D. In this section, the question addressed is: 'Is the exposure E associated with the disease D?' This is done by testing the null hypothesis of no association between E and D. Two significance tests are described — the chi-squared test for no association and Fisher's exact test. In Subsection 4.1, the test statistic for the chi-squared test is developed; and in Subsection 4.2, the test is described. Fisher's exact test is discussed briefly in Subsection 4.3.

4.1 The chi-squared test statistic

In medical statistics, it is common to report the results of a significance test for no association as well as quoting the estimated odds ratio or the relative risk and a confidence interval.

In a significance test, the evidence against a specified null hypothesis is quantified using a p value. This is the probability that data at least as extreme as those observed would have arisen if the null hypothesis were true. In the context of the cohort studies and case-control studies considered so far, the null hypothesis is that there is no association — that is, the odds ratio OR is equal to 1. Thus a significance test quantifies the evidence against the null hypothesis of no association.

If there are more than two exposure categories then, as you saw in Subsection 3.3, the strength of association cannot be summarized by a single odds ratio. One approach is to undertake an overall significance test for no association, before investigating the data in more detail.

Example 4.1 Childhood asthma and gestational age

The proportion of people with asthma is increasing in many parts of the world, though there are large geographical variations. The reasons for the increase, and for the variations in asthma rates, are not known. It has been suggested that environmental factors operating at any time after conception could be the cause. For example, it has been suggested that factors related to the development of the baby before birth may be involved. This hypothesis was investigated in a large cohort study in Denmark.

Pregnant women in Odense and Aalborg were recruited and their babies were followed up for asthma up to age 12 years. The disease D was hospitalization for asthma. The investigators analysed several variables related to foetal growth. In this example, gestational age (that is, the duration of pregnancy) is considered. Births are classified as Pre-term (the baby was premature), Term (the baby was born close to its due date) or Post-term (the baby was overdue). The data are in Table 4.1.

Yuan, W., Basso, O., Sorensen, H.T. and Olsen, J. (2002) Fetal growth and hospitalization with asthma during early childhood: a follow-up study in Denmark. *International Journal of Epidemiology*, **31**, 1240–1245.

Table 4.1 Gestational age and childhood asthma

Gestational age	Hospitalized for asthma	Not hospitalized for asthma	Total
Pre-term	18	266	284
Term	402	8565	8967
Post-term	45	1100	1145

One approach to analysing these data is to select one group as a reference category — for example, the Term group. Then relative risks or odds ratios relative to the reference category can be estimated, and 95% confidence intervals calculated. This is the approach that was described in Subsection 3.3. For the data in Table 4.1, estimates of the relative risks are

$$\widehat{RR}_{\text{pre}} = \frac{18/284}{402/8967} \simeq 1.41, \quad \widehat{RR}_{\text{post}} = \frac{45/1145}{402/8967} \simeq 0.88.$$

The 95% confidence intervals are $(0.89, 2.23)$ for RR_{pre} and $(0.65, 1.19)$ for RR_{post}. However, if the Post-term category had been chosen as reference, then the estimated relative risk $\widehat{RR}_{\text{pre}}$ would be given by

$$\widehat{RR}_{\text{pre}} = \frac{18/284}{45/1145} \simeq 1.61,$$

and the 95% confidence interval for RR_{pre} would be $(0.95, 2.74)$. Thus a different choice of reference category produces different results, as would be expected. One advantage of an overall test for no association is that it does not require a reference category to be chosen. ◆

In testing the null hypothesis of no association, the observed frequencies are compared with the frequencies that would be expected if there were no association. So the first step is to calculate these expected frequencies. The calculation is done conditional on the row totals and column totals for the observed frequencies. Table 4.2 shows the asthma data from Table 4.1 with the column totals added.

Table 4.2 Gestational age and childhood asthma with marginal totals

Gestational age	Hospitalized for asthma	Not hospitalized for for asthma	Total
Pre-term	18	266	284
Term	402	8 565	8 967
Post-term	45	1 100	1 145
Total	465	9 931	10 396

The row totals and column totals are collectively called **marginal totals**. Also shown in the bottom right-hand corner of Table 4.2 is the overall total: 10 396.

Under the null hypothesis of no association between gestational age and asthma, the probability of being hospitalized for asthma is the same in each group (Pre-term, Term and Post-term). Conditional on the marginal totals given in Table 4.2, this probability is 465/10 396. We would expect this proportion of the 284 children who were pre-term babies to be hospitalized for asthma. So the expected frequency of pre-term children hospitalized for asthma is

$$\frac{465}{10\,396} \times 284 \simeq 12.70.$$

Similarly, the expected frequency of term children not hospitalized for asthma is obtained by multiplying the overall proportion not hospitalized (9931/10 396) by the number in the term group (8967); this gives

$$\frac{9931}{10\,396} \times 8967 \simeq 8565.92.$$

In general, the expected frequency in any given cell under the hypothesis of no association is obtained using the following formula:

$$\text{expected frequency} = \frac{\text{row total} \times \text{column total}}{\text{overall total}}. \tag{4.1}$$

Note that the expected frequency is generally not an integer.

Example 4.2 Calculation of expected frequencies under the hypothesis of no association

Applying Formula (4.1) to the data in Table 4.2 leads to the expected frequencies shown in Table 4.3.

Table 4.3 Expected frequencies for childhood asthma data

Gestational age	Hospitalized for asthma	Not hospitalized for asthma	Total
Pre-term	12.70	271.30	284
Term	401.08	8565.92	8 967
Post-term	51.21	1093.79	1 145
Total	464.99	9931.01	10 396

Notice that the entries in the column labelled 'Not hospitalized for asthma' in Table 4.3 sum to 9931.01, not 9931 (as in Table 4.2); and the entries in the column 'Hospitalized for asthma' sum to 464.99, not 465 (as in Table 4.2). These discrepancies are due to rounding errors; they are not important and may be ignored.

Overall, the differences between the observed frequencies in Table 4.2 and the expected frequencies in Table 4.3 appear to be quite small. ♦

Examples 4.1 and 4.2 used data from a cohort study. The method for obtaining the expected frequencies under the null hypothesis of no association is exactly the same for case-control studies as for cohort studies.

Activity 4.1 Sleeping position and SIDS

In Example 3.4, odds ratios and confidence intervals were calculated for data on sleeping position and sudden infant death syndrome (SIDS). The data from Table 3.7 are reproduced in Table 4.4. The row totals and the overall total are also given.

Table 4.4 Sleeping position and deaths from SIDS

Position baby last placed down to sleep	Cases	Controls	Total
On its front	30	24	54
On its side	76	241	317
On its back	82	509	591
Total	188	774	962

(a) Obtain the expected frequencies under the null hypothesis of no association between sleeping position and SIDS.

(b) In which cells are the observed numbers of cases greater than expected? What does this suggest about a possible association between sleeping position and SIDS?

In Example 4.2, the observed frequencies were quite close to those expected. In Activity 4.1, the differences were larger; for example, 30 cases were last placed down on their front, compared to 10.55 expected if there were no association. This might suggest that the null hypothesis of no association should be rejected. However, since the differences might be due to random variation, a formal significance test is required before a conclusion can be drawn.

Consider data from a cohort study or a case-control study, arranged in a table with r rows (usually exposure groups) and c columns (usually disease outcome groups). Thus there are $r \times c$ cells in the table, not counting the margins. Let O_i denote the *observed* count in the ith cell, and let E_i denote the *expected* frequency for that cell, calculated under the null hypothesis of no association. The first step to constructing the significance test is to decide upon a test statistic. This should reflect the magnitudes of the differences $O_i - E_i$. A convenient test statistic is

$$\chi^2 = \sum_i \frac{(O_i - E_i)^2}{E_i}.$$

The symbol χ is the Greek letter chi. It is pronounced 'kye'.

This statistic is called the **chi-squared test statistic**. It was devised by Karl Pearson (Figure 4.1), so it is sometimes called the Pearson chi-squared statistic. It is the basis of a commonly used significance test known as the **chi-squared test** for no association. If the value of χ^2 is 0, this indicates perfect agreement between the observed frequencies and the expected frequencies. In general, the greater the differences are between the observed frequencies and those expected under the null hypothesis of no association, the larger is the value of χ^2.

Figure 4.1 Karl Pearson (1857–1936)
© Science Photo Library

Example 4.3 *The chi-squared test statistic for the childhood asthma data*

In Example 4.1, data were presented from a cohort study on the relationship between gestational age (categorized as Pre-term, Term and Post-term) and hospitalization for asthma during childhood. The expected frequencies under the hypothesis of no association were calculated in Example 4.2. The observed frequencies and the expected frequencies are shown together in Table 4.5 (the expected frequencies are in brackets).

Table 4.5 Observed frequencies O_i and expected frequencies E_i (in brackets) for childhood asthma data

Gestational age	Hospitalized for asthma	Not hospitalized for asthma	Total
Pre-term	18 (12.70)	266 (271.30)	284
Term	402 (401.08)	8565 (8565.92)	8967
Post-term	45 (51.21)	1100 (1093.79)	1145

So the value of the chi-squared test statistic is given by

$$\chi^2 = \sum_i \frac{(O_i - E_i)^2}{E_i}$$

$$\simeq \frac{(18 - 12.70)^2}{12.70} + \frac{(402 - 401.08)^2}{401.08} + \frac{(45 - 51.21)^2}{51.21}$$

$$+ \frac{(266 - 271.30)^2}{271.30} + \frac{(8565 - 8565.92)^2}{8565.92} + \frac{(1100 - 1093.79)^2}{1093.79}$$

$$\simeq 2.2118 + 0.0021 + 0.7531 + 0.1035 + 0.0001 + 0.0353$$

$$\simeq 3.11.$$

In this calculation the expected frequencies were rounded to two decimal places. If full accuracy is retained for the expected frequencies, then the value obtained for the chi-squared test statistic is 3.104. ♦

Activity 4.2 *The chi-squared test statistic for the SIDS data*

Using the expected frequencies that you calculated in Activity 4.1, obtain the value of the chi-squared test statistic for the data on sleeping position and SIDS in Table 4.4.

4.2 The chi-squared test for no association

In Subsection 4.1, the test statistic for the chi-squared test for no association was developed. In order to complete the test, the null distribution of the test statistic, that is its distribution under the null hypothesis, must be known. In this subsection, the null distribution is stated without proof and the test is described.

The null distribution of the chi-squared test statistic may be approximated by a standard continuous probability distribution known as a **chi-squared distribution**. This is a member of the chi-squared family of distributions, which is indexed by a parameter ν called the **degrees of freedom**; ν takes the values $1, 2, 3, \ldots$. The chi-squared distribution with ν degrees of freedom is denoted $\chi^2(\nu)$.

The probability density functions (p.d.f.s) for chi-squared distributions with 1, 2, 4 and 8 degrees of freedom are shown in Figure 4.2.

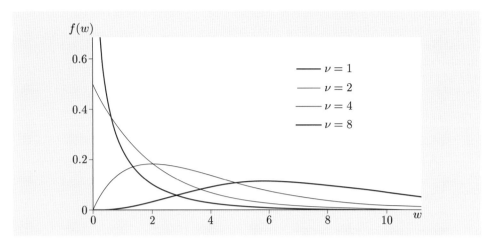

Figure 4.2 The probability density functions for four chi-squared distributions

The larger the value of the degrees of freedom parameter ν, the larger is the mean of the distribution and the more spread out the distribution. This is apparent for the four distributions represented in Figure 4.2. In fact, a random variable W with the chi-squared distribution on ν degrees of freedom has mean ν and variance 2ν. Note also that a chi-squared random variable W is defined for strictly positive values only.

The p.d.f. of the random variable $W \sim \chi^2(\nu)$ is rather complicated and is not given here. It is customary to use a computer or a table to obtain quantiles of $\chi^2(\nu)$. A table of quantiles for a range of values of ν is given in the *Handbook*. Part of the table is reproduced in Table 4.6.

Table 4.6 Selected quantiles for chi-squared distributions

ν	0.80	0.90	0.95	0.975	0.99	0.995	0.999
1	1.64	2.71	3.84	5.02	6.63	7.88	10.83
2	3.22	4.61	5.99	7.38	9.21	10.60	13.82
3	4.64	6.25	7.81	9.35	11.34	12.84	16.27
4	5.99	7.78	9.49	11.14	13.28	14.86	18.47

Example 4.4 Using the table

The 0.95-quantile of $\chi^2(1)$ is the number in Table 4.6 in the row labelled 1 (corresponding to $\nu = 1$) and in the column headed 0.95. This is 3.84. Similarly, the 0.99-quantile of $\chi^2(3)$ is 11.34. ♦

Activity 4.3 Tail probabilities for chi-squared distributions

Use the table of quantiles for chi-squared distributions in the *Handbook* to answer the following questions.

(a) Find the 0.95-quantile of $\chi^2(5)$.

(b) Find the value w such that $P(W > w) = 0.025$, where $W \sim \chi^2(8)$.

(c) Find the best possible lower bound and the best possible upper bound available from the tables for $P(W > 10.25)$, where $W \sim \chi^2(3)$.

Section 4 Testing for no association in cohort studies and case-control studies

A table with r rows and c columns (excluding row totals and column totals) is called an $r \times c$ table; this is read 'r by c'. For instance, Table 4.5 is a 3×2 table. The value of the degrees of freedom parameter ν for the chi-squared test for no association in such a table is $(r-1) \times (c-1)$.

For example, for a 3×2 table, that is, one with three rows and two columns (so that $r = 3$ and $c = 2$), the degrees of freedom are $\nu = (3-1) \times (2-1) = 2$.

The degrees of freedom have a direct interpretation in terms of the maximum number of cells that can be specified freely, subject to the constraints imposed by the marginal totals. This is illustrated in Example 4.5.

Example 4.5 Interpretation of the degrees of freedom

Table 4.4 has three rows and two columns: $r = 3$ and $c = 2$. The layout of the table, emptied of its contents except for the marginal totals, is shown in Figure 4.3(a).

Now try and enter numbers into the six cells of the table in such a way that the row totals and column totals are respected. For example, start with the number 50 in the top left-hand cell (you have some freedom of choice here). The value in the cell in row 1, column 2 must then be 4, since the values in row 1 must add up to 54. The number 50 was a free choice and is entered in bold in Figure 4.3(b); the number 4 was not, and is entered in italics.

Moving to row 2, enter a number in the first column. (You also have some freedom in selecting this value.) Suppose that you pick the number 100. Since the numbers in row 2 must sum to 317, the value in column 2 must be 217. These values are shown in Figure 4.3(c).

But now, note that the remaining two empty cells in the table are also determined: since the numbers in column 1 sum to 188, the remaining entry in column 1 must be 38; and, similarly, the remaining entry in column 2 must be 553. (See Figure 4.3(d).)

Thus you are able to exercise some choice in filling only two of the six cells of the table. Another way of describing this fact is that, conditional on the marginal totals, the table has two degrees of freedom.

Similarly, for a 3×3 table, there are $(3-1) \times (3-1) = 4$ degrees of freedom, meaning that you have some discretion in filling in four of the nine cells of the table, the values in the remaining five cells then being wholly determined by the marginal totals. ◆

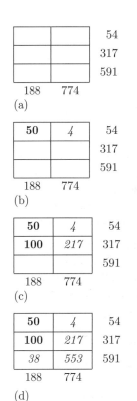

Figure 4.3 Filling in the cells

Return now to the chi-squared test for no association. You have seen that the null distribution of the test statistic is approximately $\chi^2(\nu)$. This approximation is adequate provided that all the expected frequencies E_i are at least 5. If this is the case, then the chi-squared distribution may be used to calculate the significance probability, or p value, for the test. The chi-squared test statistic measures the extent to which observed frequencies differ from those expected under the hypothesis of no association: the higher the value of χ^2, the greater the discrepancy between the observed and expected frequencies. Thus the test is one-sided: only high values of χ^2 provide evidence against the hypothesis of no association. So only the upper tail of $\chi^2(\nu)$ is used in calculating the p value.

The p value can be calculated using a computer, or tables can be used to obtain an approximate value for p as in part (c) of Activity 4.3. Small p values suggest that data as extreme as those observed are unlikely to have arisen by chance if the null hypothesis of no association is true. So a small p value is interpreted as evidence against the null hypothesis of no association.

Example 4.6 Sleeping position and SIDS: calculating the p value

In Activity 4.2, you found that, for the data in Table 4.4 on SIDS and sleeping position, the value of the chi-squared test statistic χ^2 is 60.60. Since the data form a 3×2 table, the null distribution of the test statistic is approximately $\chi^2(2)$. The expected frequencies are all at least 5, so the approximation is adequate.

From Table 4.6, the 0.999-quantile of $\chi^2(2)$ is 13.82. Since the observed value of the test statistic, 60.60, is greater than this, it follows that the p value of the test for no association is less than $0.001 : p < 0.001$. In fact, using a computer gives a p value of 6.93×10^{-14}; such a tiny value would usually be reported as $p < 0.0001$.

This small p value indicates that data as extreme as those observed are very unlikely to have arisen by chance if the null hypothesis of no association were true. This is strong evidence that the null hypothesis is not true. The conclusion is that there is a statistically significant association between sleeping position and SIDS. ♦

In the context of cohort studies and case-control studies, the p value can be interpreted in terms of evidence against the null hypothesis of no association. For example, an estimated odds ratio greater than 1 or less than 1 provides *some* evidence of an association (only if it were exactly equal to 1 could it truly be said to provide *no* evidence of an association). However, if the p value of the test for no association is large (greater than 0.1, say), then there is little evidence against the null hypothesis of no association from this particular study, and hence little evidence of an association. A rough guide to the interpretation of p values in the context of cohort studies and case-control studies is provided in Table 4.7.

Table 4.7 Interpretation of p values for the test for no association

Significance probability p	Rough interpretation
$p > 0.10$	little evidence of an association
$0.10 \geq p > 0.05$	weak evidence of an association
$0.05 \geq p > 0.01$	moderate evidence of an association
$p \leq 0.01$	strong evidence of an association

The thresholds in Table 4.7 are to some extent arbitrary. For example, p values of 0.049 and 0.051 should lead to broadly similar conclusions.

Activity 4.4 Childhood asthma: calculating and interpreting the p value

In Example 4.3, the value of the chi-squared test statistic for the test for no association between gestational age and childhood asthma was found to be 3.11.

(a) Obtain a range of values for the p value for the test for no association between gestational age and asthma.

(b) Interpret your results.

Section 4 Testing for no association in cohort studies and case-control studies

The procedure for the chi-squared test for no association between the variables in an $r \times c$ table is summarized in the following box.

> **The chi-squared test for no association in an $r \times c$ table**
> 1. Calculate a table of expected frequencies under the null hypothesis of no association: the expected frequency for a cell is given by
> $$\text{expected frequency} = \frac{\text{row total} \times \text{column total}}{\text{overall total}}.$$
> 2. Calculate the value of the chi-squared test statistic using the formula
> $$\chi^2 = \sum_i \frac{(O_i - E_i)^2}{E_i},$$
> where O_i is the observed frequency and E_i is the expected frequency for the ith cell, and where the summation is over the $r \times c$ cells in the table.
> 3. Obtain the null distribution of the test statistic: under the null hypothesis of no association, the distribution of the test statistic is approximately chi-squared with $\nu = (r-1) \times (c-1)$ degrees of freedom.
>
>   The approximation is adequate provided that all the expected frequencies are at least 5.
> 4. Calculate the p value for the test and interpret your answer.

Activity 4.5 Ectopic pregnancy and history of genital infections

In Activity 3.3, you calculated odds ratios and confidence intervals using data on ectopic pregnancy and history of genital infections. These data are reproduced in Table 4.8. Carry out the chi-squared test for no association using these data.

Table 4.8 Ectopic pregnancy and history of genital infections

History	Cases	Controls
PID	212	112
Non-PID	157	407
None	411	1154
Total	780	1673

Note that the *strength of evidence* of an association (as quantified by a p value) is not the same thing as the *strength of the association* (as quantified by the odds ratio). For example, a large study can provide strong evidence of a weak association: the p value may be very small, though the odds ratio is close to unity. In medical statistics, both strength of evidence and strength of association are of interest, so it is usual to carry out a significance test and also to quote the odds ratio or relative risk and a confidence interval.

Activity 4.6 will give you some practice in carrying out the chi-squared test for no association, calculating confidence intervals, and reporting the results.

Activity 4.6 Smoking and lung cancer

In Example 3.1, data from the historic 1950 case-control study of smoking and lung cancer by Doll and Hill were presented. The data are reproduced in Table 4.9.

Table 4.9 Smoking and lung cancer in males

Exposure category	Cases of lung cancer	Controls
Smoked	647	622
Never smoked	2	27
Total	649	649

(a) Test the null hypothesis that smoking and lung cancer in males are not associated.
(b) Estimate the odds ratio for smoking and lung cancer in males, and obtain a 95% confidence interval for the odds ratio.
(c) Interpret your results.

4.3 Fisher's exact test

When no expected frequencies are less than 5, the null distribution of the chi-squared test statistic is well approximated by a chi-squared distribution, and the chi-squared test gives reliable results. However, when one or more expected frequencies are less than 5, the chi-squared test may give unreliable results. Such a situation is described in Example 4.7.

Example 4.7 Measles case fatality rate in South Africa

In South Africa, measles vaccination campaigns were conducted in 1996 and 1997 with the aim of stopping the circulation of the measles virus. To evaluate the campaign, data on measles were collected corresponding to the periods before and after the campaign.

The mass campaign very substantially reduced the incidence of measles. In the Western Cape Province, there were 736 cases of measles hospitalized in the period 1992–97, prior to the campaign, of whom 23 died. Between the beginning of 1998 and July 1999, after the campaign, there were only 29 cases of measles sufficiently serious to be hospitalized, none of whom died. The data are shown in Table 4.10.

Uzicanin, A., Eggers, R., Webb, E. *et al.* (2002) Impact of the 1996–1997 supplementary measles vaccination campaigns in South Africa. *International Journal of Epidemiology*, **31**, 968–976.

Table 4.10 Measles hospitalizations and deaths

Period	Died	Did not die	Total
After campaign	0	29	29
Before campaign	23	713	736
Total	23	742	765

The number expected to die after the campaign under the hypothesis of no association between the measles vaccination campaign and the proportion of serious cases who died is, using (4.1),

$$\frac{29 \times 23}{765} \simeq 0.87.$$

So the observed number, zero, is not very different from that expected: the vaccination campaign reduced the number of measles cases, but probably did not affect the proportion of cases who died. But how can this be tested formally? The chi-squared test for no association may not be valid in this instance as the expected frequency for the zero cell is less than 5. ♦

In fact, there is an exact test that applies in all circumstances. This is **Fisher's exact test**, named after the renowned statistician Ronald Aylmer Fisher (Figure 4.4). The test involves working through all possible tables with the same marginal totals as the table observed, and summing the probabilities of those that are as extreme or more extreme than the observed table. This is a time-consuming procedure, and is best done by a computer. Fisher originally developed the test for use with 2 × 2 tables, and it was later generalized to tables of arbitrary dimensions. For large tables, the computations can be extremely demanding, even with the use of a computer. So application of the test will be left to the computer book.

Figure 4.4 Ronald Aylmer Fisher (1890–1962)
© Science Photo Library

Summary of Section 4

In this section, the chi-squared test of the null hypothesis of no association has been described in the context of cohort studies and case-control studies. The chi-squared test statistic is a measure of discrepancy between the observed frequencies and the frequencies expected under the null hypothesis. The null distribution is approximately a chi-squared distribution. The approximation is good and the chi-squared test gives reliable results when none of the expected frequencies is less than 5. In other cases, Fisher's exact test may be used.

Exercise on Section 4

Exercise 4.1 Seat belts and children's safety in car accidents

In Activity 1.2, data on seat belt use and injuries sustained by children in car accidents were described. The data are reproduced in Table 4.11.

Table 4.11   Seat belt use and injury sustained by children aged 4–14

Exposure category	D: sustained at least moderately severe injury		
	Yes	No	Total
Not wearing a seat belt (E)	14	19	33
Wearing a seat belt (not E)	13	39	52

Halman, I., Chipman, M., Parkin, P.C. and Wright, J.G. (2002) Are seat belt restraints as effective in school age children as in adults? A prospective crash study. *British Medical Journal*, **324**, 1123–1125.

Carry out a chi-squared test for no association between children sustaining at least moderately severe injury and failing to wear a seat belt.

5 Analysing cohort studies and case-control studies in SPSS

In this section, you will learn how to use SPSS to do calculations of the type described in Sections 1 to 4.

Refer to Chapters 1 and 2 of Computer Book 1 for the work in this section.

Summary of Section 5

Entering tabular data in SPSS and creating tables has been described. The facilities in SPSS for analysing data from cohort studies and case-control studies have been explored. These include estimating odds ratios and (for cohort studies) relative risks, calculating confidence intervals, obtaining expected frequencies under the null hypothesis of no association, and testing for no association using the chi-squared test and Fisher's exact test.

Part II Bias, confounding and causation

Introduction to Part II

In Part I, you learned about two types of studies, the cohort study and the case-control study. These types of studies are commonly used in epidemiology to determine whether an exposure E is associated with a disease D, and, if so, how strong the association is.

Like other types of evidence, statistical evidence for an association (or lack of evidence of association) should be subjected to scrutiny to decide whether it is trustworthy. Thus the first question a statistician should ask when presented with evidence of an association is whether the association is genuine, or whether it could be an artefact resulting from the particular way the data were collected. This process of criticism is central to medical statistics, and indeed to evaluating statistical evidence in other fields. In Part II, you will learn some of the statistical concepts and methods involved. In Section 6, various ways in which spurious associations may be generated or genuine associations concealed are discussed. Then, in Section 7, you will learn how to adjust your analyses for sources of variation that may distort associations.

Now suppose that you have established that, as far as you can tell, an association is genuine, or at least that it is not obviously spurious. What next? You may remember that, at the start of Part I, you were reminded of an important limitation of statistical analyses: *association does not imply causation*. In other words, an exposure and a disease may be associated, without the exposure causing the disease. However, what we really want to know is whether E causes D: whether smoking *causes* lung cancer, whether hormone-replacement therapy *causes* breast cancer, whether vaccines *protect* against infection and whether they *cause* adverse events. But if causation cannot be deduced from association, how can it be established?

There is no purely statistical approach to demonstrating that an association is causal. However, there are some statistical tools that can be used to evaluate evidence for causation, and one of these is discussed in Section 8. Finally, in Section 9, you will learn how the methods introduced in Sections 6 to 8 are implemented in SPSS.

6 Bias and confounding

In this section, you will learn about various biases that may distort or conceal an association (or lack of association) between an exposure E and a disease D. Most studies include a discussion of possible sources of bias and their likely impact on the results. This aspect of the statistical analysis is often qualitative: it contrasts with the more formal statistical approach involving p values and confidence intervals. However, the evaluation of potential biases forms an integral part of the statistical analysis, and its importance cannot be overstated. Subsection 6.1 defines what is meant by bias in the context of medical statistics. In Subsections 6.2 to 6.4, three types of bias are discussed.

Section 6 Bias and confounding

6.1 What is bias?

A study is said to be **biased** if some aspect of the design, sampling, data collection or analysis method produces results that systematically overestimate or underestimate the strength of association. The word *systematic* is the important one here: bias is not the same as sampling error. Sampling error is the result of random variation. If an unbiased study were repeated many times, the estimates $\widehat{OR}$ of the odds ratio would vary randomly around their underlying value OR. Bias, on the other hand, results in estimates $\widehat{OR}$ that are systematically lower or higher than OR, because they are estimating some quantity OR^* which differs from the true value OR. The **bias** is the difference $OR^* - OR$. Sampling error may be reduced by increasing the sample size, and can be quantified using statistical theory (for example, with confidence intervals). In contrast, biases cannot be reduced in this way: increasing the sample size just produces a better estimate of the wrong parameter OR^*. The way to control bias is to improve the design or adjust the analysis.

You may have come across the word 'bias' in connection with estimators: an estimator $\widehat{\theta}$ of a parameter θ is said to be unbiased if $E(\widehat{\theta}) = \theta$, otherwise it is biased. In this book, the word bias is used in a different sense: it relates to the study as a whole, not to an estimator.

Choosing the sample size is discussed in Section 11.

Example 6.1 Measuring temperatures

Suppose that two nurses A and B take the temperatures of the same patients at roughly the same times. If the measurements taken by nurse A are unbiased, they will broadly agree with the patients' true temperatures. There will be differences on individual patients, owing for example to random measurement errors. But these differences are not systematic, and will cancel out on average.

Suppose now that the measurements taken by nurse B are biased — caused by a defective thermometer, for example, or by poor technique on the part of the nurse. Suppose that the nurse produces values that are systematically lower than the true temperatures. This bias will not cancel out by averaging over all the patients. For nurse A, the difference between the measured value and the true value could be represented by the normal distribution $N(0, \sigma^2)$, where σ^2 represents the variance of the random measurement errors. For nurse B, on the other hand, the difference between the measured value and the true value is $N(\theta, \sigma^2)$, with $\theta < 0$. Taking more measurements will not remedy the fact that $\theta < 0$. ♦

The most obvious cause of bias in comparative studies such as cohort studies and case-control studies is failure to compare like with like. Bias due to lack of comparability can arise in many different guises. In Subsections 6.2 to 6.4, you will learn about three important types of bias: *selection bias*, *information bias* and *confounding*.

6.2 Selection bias

As its name implies, **selection bias** is bias that occurs in connection with the selection of individuals for study: it occurs if the way in which individuals are selected induces differences between the two groups to be compared (exposed and unexposed in cohort studies, cases and controls in case-control studies). Selection bias might occur, for example, if the procedures used for selecting participants are different in different groups. Selection bias can also occur unwittingly: even when the procedures are the same for all, it can occur simply because individuals differ in their probability of being included in the registries, hospital admission records, patient lists and other databases that are used to select participants.

Example 6.2 Measles vaccine and inflammatory bowel disease

In 1995 a cohort study was published purporting to show that measles vaccination was positively associated with various forms of inflammatory bowel disease (IBD), including Crohn's disease (relative risk $RR = 3.01$, 95% confidence interval 1.45 to 6.23) and ulcerative colitis ($RR = 2.53$, 95% confidence interval 1.15 to 5.58).

Thompson, N.P., Montgomery, S.M., Pounder, R.E. and Wakefield, A.J. (1995) Is measles vaccination a risk factor for inflammatory bowel disease? *Lancet*, **345**, 1071–1074.

The study compared the risks of IBD in two groups. The exposed group comprised 3545 adults who received single measles vaccine in childhood. The unexposed group comprised 11 407 adults who did not receive measles vaccine. The data from the study are in Table 6.1.

Table 6.1 Measles vaccination and inflammatory bowel disease

	Crohn's disease	Ulcerative colitis	Total in cohort
Vaccinated	14	11	3 545
Unvaccinated	15	14	11 407

The exposed group of 3545 were part of a larger group of 9577 vaccinated children who took part in a study undertaken by the Medical Research Council in 1964. After the study, the participants were followed up at regular intervals using health questionnaires. Those still answering questionnaires in 1994 formed the exposed group in the study. Thus 37% (3545/9577) of the original cohort were eventually selected for inclusion in the 1995 study.

The unexposed group of 11 407 were part of a larger group of 17 414 recruited to the 1958 National Child Development Study. These participants were traced over the years, using a range of active search methods. Thus 66% (11 407/17 414) of the original cohort were eventually selected for inclusion in the 1995 study.

The selection procedure for the two cohorts thus differed in the intensity of follow-up: 37% for the exposed group, 66% for the unexposed group.

Could the difference result in selection bias? At this point we have to use our judgement: a definitive answer is not possible. It is possible, even likely, that people with chronic illnesses such as inflammatory bowel disease will be more likely to answer health questionnaires than people without chronic diseases. If this is true, then the 37% of remaining participants in the vaccinated cohort may include a higher proportion of patients suffering IBD than the 66% in the unvaccinated cohort. This would produce a spurious positive association. ♦

In Example 6.2 the selection procedure was shown to differ for the exposed and unexposed groups. This suggests that the study may be prone to selection bias. However, bias has not been demonstrated: the most that can be done is to examine the study design critically and come to a judgement about the presence or otherwise of selection bias.

Case-control studies are particularly prone to selection bias. This is because it may be difficult, and sometimes wholly impossible, to select the controls using the same procedure and from the same population as the cases.

Section 6 Bias and confounding

Example 6.3 Political activity and homicide in Karachi

In Activity 3.1, a case-control study to investigate the evidence for an association between political activity and homicide in a district of Karachi, Pakistan, was described. The cases included 35 homicide victims. There were 85 controls, selected from comparable families. The exposure of interest was attendance at political meetings. The data are shown in Table 6.2.

Table 6.2 Homicide and political activity in Karachi

Exposure category	Cases	Controls
Attended political meetings	11	2
Did not attend political meetings	24	83
Total	35	85

The odds ratio for attendance at political meetings is 19.02, with 95% confidence interval (3.94, 91.76), indicating a strong positive association between attendance at political meetings and death by homicide.

Could this association be due to selection bias? To answer this we need to examine closely the way the cases and controls were selected. The cases were identified with the help of community organizations working in the study area: field workers for these organizations directly identified households where they knew that someone had been killed. The control households were selected using a selection procedure involving an element of randomness: a point was chosen at random on a map and neighbouring houses were identified. If a household agreed to participate, a control individual was chosen within the household in the 18–60 age group. (Further selection criteria were defined in case there were several potential controls.)

Clearly, the selection procedure differs for cases and controls. Could this lead to selection bias, and, if so, would the bias be large enough to affect the results substantially? A particular concern is that households with politically active members may be more likely to be known to community organizations than those without politically active members. If this is so, this may lead to politically active homicide victims being over-represented among the cases, simply by virtue of the case selection method. This would bias the odds ratio upwards. ♦

Examples 6.2 and 6.3 illustrate the fact that it is often not possible to *prove* that selection bias occurred. Rather, it is a matter of applying common sense, and coming to a judgement about how reliable the results may be.

Activity 6.1 Smoking and lung cancer

In Example 3.1, data were presented from a famous case-control study of smoking and lung cancer by Doll and Hill. The data for males are shown in Table 6.3.

Table 6.3 Smoking and lung cancer in males

Exposure category	Cases of lung cancer	Controls
Smoked	647	622
Never smoked	2	27
Total	649	649

The 649 cases were male patients admitted to hospital with lung cancer. For each case, a control was selected of the same sex and similar age, who was admitted to the same hospital for a disease other than cancer; if there were several eligible controls, the first on the ward list was chosen. The odds ratio for smoking is 14.04, with 95% confidence interval (3.33, 59.31).

(a) From the description given, identify any differences between the selection procedures for cases and controls.

(b) In your view, could these differences or any other selection mechanism account for the difference between the proportions of smokers among the cases and controls?

(c) The controls were selected from hospital patients with a disease other than cancer. It has subsequently been shown that smoking causes diseases other than cancer — heart disease, for example. In the light of this later finding, what effect do you think this choice of controls might have on the odds ratio?

6.3 Information bias

Suppose that the individuals to be studied have been selected. A further bias, called **information bias**, may arise through the process of gathering information on these individuals. This bias may be due to differences in data collection or measurement methods, or differences in the quality of responses obtained, between the two groups to be compared.

For example, when collecting data on variables such as disease outcomes in cohort studies or exposure status in case-control studies, errors might occur. Thus, a disease may be misdiagnosed, leading to a case of disease being misclassified as not having the disease. Similarly, an exposed individual may be misclassified as unexposed, and vice versa. Good studies are designed to reduce such misclassification to a minimum, though some errors are inevitable. However, if misclassification occurs more frequently in one group than another, this can bias the results seriously.

Example 6.4 Measles vaccine and inflammatory bowel disease

In Example 6.2, a cohort study of a putative association between single measles vaccination and inflammatory bowel disease was described.

The data from the study are in Table 6.1.

The vaccinated individuals were sent a questionnaire specifically asking whether they had Crohn's disease, ulcerative colitis or other types of inflammatory bowel disease (IBD). In contrast, information from the unvaccinated individuals, who were part of the 1958 National Child Development Study, was obtained from the standard questionnaire for this study in which they were asked about any long-standing illness and contacts with health services.

Could the study be affected by information bias? The methods used for classifying individuals as having IBD or not having IBD differ according to whether the individuals are vaccinated or not. In particular, the specific questions about IBD to the vaccinated individuals contrast with the more general questions to the unvaccinated individuals. More specific questioning is likely to elicit a more accurate response. Thus information bias may have occurred in this study, and it is perhaps likely to bias the relative risk upwards. ◆

As with selection bias, identifying possible sources of information bias requires common sense, and is helped by knowledge of the context of the study.

Activity 6.2 Political activity and homicide in Karachi

A case-control study to investigate the association between attendance at political meetings and death by homicide in a district of Karachi, Pakistan was discussed in Example 6.3.

The data are in Table 6.2.

Since the cases had been killed, the questionnaire obviously could not be administered to them. Instead, it was administered to a close relative (usually the victim's wife). Since even a close relative may not know of the case's political activities, information on the controls was obtained in a similar indirect manner as for the cases. Thus, the questionnaire was administered to a close relative of the control in each case, rather than to the control directly.

The question of primary interest relates to the attendance of the study subject (case or control) at political meetings. In your view, are the data collected on this likely to suffer from information bias? In formulating your answer, consider the following aspects.

(a) Was the information on attendance at political meetings collected in the same way for cases as for controls?

(b) Could the accuracy of the data on attendance at political meetings of cases and controls have been different?

(c) If information bias did arise, would it have operated so as to increase or decrease the odds ratio?

6.4 Confounding

There is very little, if anything, that can be done to correct for selection bias and information bias once they have occurred. In this subsection, another type of bias is discussed, resulting from a phenomenon called **confounding**. This bias can be corrected in some circumstances.

Confounding bias differs from selection bias and information bias in that it involves a third variable, which may distort the association between the exposure and the disease. Confounding can have spectacularly disastrous results, and it is perhaps the single most important type of bias in medical statistics.

Example 6.5 Treatment for kidney stones

Kidney stones can cause intense pain, obstruct the urinary tract or damage the kidneys. In the 1980s several new techniques were introduced to remove kidney stones, in preference to the then standard form of treatment which involved open surgery. These new techniques included keyhole surgery and shockwave therapy.

A study was undertaken to compare the different treatment methods. This example focuses on open surgery and keyhole surgery (the technical term is percutaneous nephrolithotomy).

Charig, R., Webb, D.R., Payne, S.R. and Wickham, J.E.A. (1986) Comparison of treatment of renal calculi by open surgery, percutaneous nephrolithotomy, and extracorporeal shockwave lithotripsy. *British Medical Journal*, **292**, 879–882.

Treatment was defined as successful if the stones were eliminated or reduced to less than 2 mm diameter after three months. The study was a cohort study, including 350 patients treated with open surgery and 350 with keyhole surgery. In this study the exposure is the new treatment, namely keyhole surgery, and the 'disease' outcomes are success or failure of the treatment. The results are shown in Table 6.4 (overleaf).

Table 6.4 Treatment of kidney stones by open surgery and keyhole surgery

Treatment	Success	Failure	Total
Keyhole surgery	289	61	350
Open surgery	273	77	350

The success rates for the two types of surgery are as follows:

keyhole surgery: $289/350 \simeq 0.8257$,

open surgery: $273/350 = 0.7800$.

Thus keyhole surgery has a better success rate than open surgery. The association between keyhole surgery and success can be quantified by the odds ratio:

$$\widehat{OR} = \frac{289 \times 77}{61 \times 273} \simeq 1.34.$$

However, suppose that we look at the data for small kidney stones and large kidney stones separately. The data for the treatment of small kidney stones, namely those less than 2 cm in diameter, are in Table 6.5.

Table 6.5 Treatment of small kidney stones (< 2 cm diameter)

Treatment	Success	Failure	Total
Keyhole surgery	234	36	270
Open surgery	81	6	87

The success rates for keyhole surgery and open surgery are now $234/270 \simeq 0.8667$ and $81/87 \simeq 0.9310$, respectively. Thus, for small stones, open surgery has a better success rate than keyhole surgery: this result is the opposite of that for all kidney stones. The corresponding odds ratio is $(234 \times 6)/(36 \times 81) \simeq 0.48$.

You might conclude that, in order to compensate for its lack of success with small stones, keyhole surgery must be much more successful than open surgery when applied to the treatment of large stones, namely those greater than or equal to 2 cm in diameter. Paradoxically, this is not the case. The data for large stones are in Table 6.6.

Table 6.6 Treatment of large kidney stones (≥ 2 cm diameter)

Treatment	Success	Failure	Total
Keyhole surgery	55	25	80
Open surgery	192	71	263

The success rates for keyhole surgery and open surgery are $55/80 = 0.6875$ and $192/263 \simeq 0.7300$, respectively. The odds ratio is again less than 1:
$\widehat{OR} = (55 \times 71)/(25 \times 192) \simeq 0.81$.

Thus, even for large stones, the success rate for keyhole surgery is less than that for open surgery. You may check that the data in Tables 6.5 and 6.6 comprise all the data in Table 6.4: none are missing. For example, the 350 stones treated with keyhole surgery include 270 small stones and 80 large ones.

The results obtained may be summarized as follows. Based on the data in Table 6.4, keyhole surgery appears to have a higher success rate than open surgery. However, for both small and large stones, keyhole surgery has a lower success rate than open surgery. So it must have a lower success rate than open surgery overall — a conclusion that contradicts that derived from the data in Table 6.4.

The conclusion must be that the odds ratio of 1.34 calculated from Table 6.4 is biased. Indeed, it is so severely biased that the direction of association has been reversed: in fact, keyhole surgery has a lower success rate than open surgery for all stone types (odds ratios of 0.48 for small stones and 0.81 for large stones)! ♦

Example 6.5 provides an instance of **Simpson's paradox**. The relationship between the exposure (treatment method) and outcome (success or failure of the treatment) is said to be **confounded** by stone size. In this example, stone size is a **confounder**.

Example 6.5 continued   Treatment for kidney stones

How did this confounding arise? The answer may be found by looking at the marginal tables of stone size by treatment method and stone size by treatment outcome. These are obtained from Tables 6.5 and 6.6, and are given in Tables 6.7 and 6.8.

Table 6.7   Stone size and treatment method

Stone size	Keyhole surgery	Open surgery
< 2 cm	270	87
≥ 2 cm	80	263

Table 6.8   Stone size and treatment outcome

Stone size	Treatment success	Treatment failure
< 2 cm	315	42
≥ 2 cm	247	96

Look first at Table 6.7. Clearly, there is a strong association between stone size and choice of treatment method: keyhole surgery was far more likely to be used than open surgery when the stone size was small, that is, for less severe cases. Similarly, in Table 6.8, stone size is associated with treatment outcome: treatment (whatever it may be) is more likely to be successful for smaller stone sizes.

Return now to Table 6.4. Table 6.7 shows that the 350 cases treated by keyhole surgery tended to be less severe than the 350 cases treated with open surgery. Since the success rate for less severe cases was higher (see Table 6.8), there was a greater proportion of successful outcomes among the keyhole surgery cases than the open surgery cases. Thus the apparent improvement in success rates with keyhole surgery, suggested by the data in Table 6.4, is simply due to the fact that fewer severe cases were treated using this method. ◆

At this point, it is convenient to introduce some new terminology. Tables 6.5 and 6.6 are said to be **stratified** by the stone size variable. This simply means that Table 6.4 has been split up into two sub-tables or **strata** determined by the values of a third variable — in this case, stone size. These values are sometimes called **levels**. Stone size is called the **stratifying variable**. The data in Table 6.4 are said to have been **aggregated** over stone size: the cell counts are just the sums of the counts in the stratified tables. Table 6.4 is called the **aggregated table**. The odds ratios for the stratified tables are called **stratum-specific** odds ratios.

Stratum is singular, strata plural.

With these definitions, you are now in a position to explore further the paradoxes of confounding.

Example 6.6   Odds ratios — now you see them, now you don't

This example of confounding is constructed using hypothetical cohort data for an exposure E and a disease D, stratified according to a third variable C that takes values (or levels) $C = 0$ and $C = 1$. The stratified data are in Table 6.9 (overleaf).

Table 6.9 Hypothetical cohort data stratified by variable C

(a) $C = 0$

	Disease	No disease
Exposed	20	2
Not exposed	60	10

(b) $C = 1$

	Disease	No disease
Exposed	60	18
Not exposed	20	10

The stratum-specific odds ratios for the association between E and D are both equal to 1.67:

$$\widehat{OR}_{C=0} = \frac{20 \times 10}{2 \times 60} \simeq 1.67, \quad \widehat{OR}_{C=1} = \frac{60 \times 10}{18 \times 20} \simeq 1.67.$$

Thus exposure is positively associated with disease at both levels of C. It may reasonably be concluded that the estimated odds ratio for the association between exposure E and disease D, irrespective of the value of the confounder C, is 1.67. This is indeed correct.

Now consider the aggregated data in Table 6.10.

Table 6.10 Data aggregated over the levels of C

	Disease	No disease
Exposed	$20 + 60 = 80$	$2 + 18 = 20$
Not exposed	$60 + 20 = 80$	$10 + 10 = 20$

Calculating the odds ratio from the aggregated table gives

$$\widehat{OR} = \frac{80 \times 20}{20 \times 80} = 1.$$

Thus aggregating the data over the levels of C conceals completely the positive association between E and D: the variable C is a confounder, and the odds ratio from the aggregated data is biased. Confounding occurs because the variable C is associated with both E and D: a greater proportion are diseased at level $C = 0$ than level $C = 1$, and a greater proportion are exposed at level $C = 1$ than at level $C = 0$. ♦

Activity 6.3 *Odds ratios — now you don't see them, now you do*

Consider the set of hypothetical cohort data in Table 6.11. As in Example 6.6, data on the association between an exposure E and a disease D are stratified according to the levels of a third variable C.

Table 6.11 Hypothetical cohort data stratified by variable C

(a) $C = 0$

	Disease	No disease
Exposed	70	30
Not exposed	14	6

(b) $C = 1$

	Disease	No disease
Exposed	14	26
Not exposed	42	78

(a) Obtain the stratum-specific odds ratios for the association between E and D for the separate levels of C. What do you conclude about the association between E and D?

(b) Form a new table by aggregating the data over the levels of C.

(c) Obtain the odds ratio for the association between E and D from the aggregated table.

(d) What do you conclude?

Section 6 Bias and confounding

The effect of confounding can be represented schematically. In Figure 6.1(a), the line between E and D represents a genuine association between E and D. In Figure 6.1(b), there is no line directly between E and D: E and D are not associated. However, in Figure 6.1(b), the confounder C is associated with both E and D, as represented by the lines connecting C to E and C to D. This confounder distorts the true relationship between E and D — for example, by inducing a spurious association between E and D. Confounding can induce a spurious association when there is no association, it can conceal a true association, it can reverse the direction of an association, and it can amplify or reduce the strength of an association without altering its direction. The nature and effect of confounding are summarized in the following box.

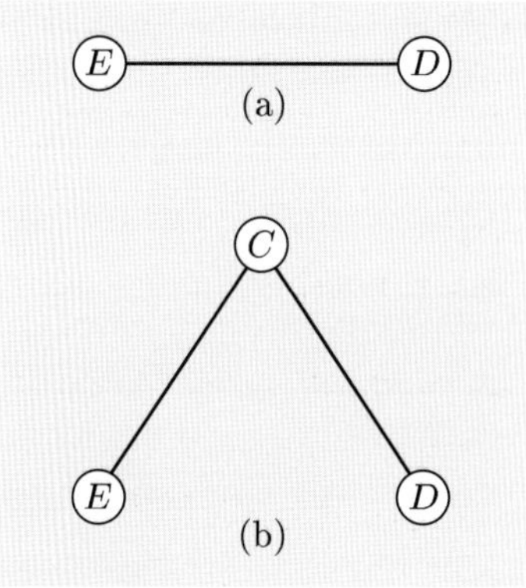

Figure 6.1 Schematic representations of (a) association and (b) confounding

> **Confounding bias**
>
> Confounding may arise if both the exposure E and the disease D are associated with a third variable C, known as a **confounder**.
>
> Aggregating data over the levels of a confounder may induce spurious associations, distort or mask true associations, or reverse the direction of associations. This effect is called **confounding bias**.

Confounding is something that medical statisticians worry about a lot. It can seriously bias the results, and goes some of the way to explaining why different studies of the same exposure and disease can produce different conclusions. Confounding can be identified by stratifying the data according to the levels of the confounder. In Section 7, you will learn how to adjust odds ratios to take account of confounding. However, stratifying the data requires data on potential confounders to be collected.

In practice, the effect of confounding is seldom as spectacular as that illustrated in Example 6.5. More commonly, confounding exaggerates or reduces the strength of an association, rather than reversing its direction. However, another striking example of confounding is provided in Activity 6.4.

Activity 6.4 Diabetes and mortality

In a cohort study of mortality in persons with diabetes, the proportions with insulin dependent and non-insulin dependent diabetes who died before the end of the study were compared. The study included 358 persons with insulin dependent diabetes and 544 with non-insulin dependent diabetes.

Julious, S.A. and Mullee, M.A. (1994) Confounding and Simpson's paradox. *British Medical Journal*, **309**, 1480–1481.

Non-insulin dependent diabetes usually develops in older people. Since death is also age-dependent, age is a possible confounder of the association (if any) between diabetes type and mortality. Accordingly, the data were stratified in two age groups: patients aged 40 years or less, and patients aged over 40 years. The stratified data are in Table 6.12, with non-insulin dependent patients as the exposed group.

Table 6.12 Diabetes and mortality

(a) Patients aged 40 or less

Diabetes type	Died	Alive	Total
Non-insulin dependent	0	15	15
Insulin dependent	1	129	130

(b) Patients aged over 40

Diabetes type	Died	Alive	Total
Non-insulin dependent	218	311	529
Insulin dependent	104	124	228

(a) Obtain the age-specific odds ratios for the association between diabetes type and mortality.

(b) Aggregate the data in Table 6.12(a) and Table 6.12(b) into a single table.

(c) Obtain the odds ratio for the aggregated data.

(d) Is non-insulin dependent diabetes associated with higher mortality than insulin dependent diabetes? Is age a confounder? What is its effect?

Summary of Section 6

A study of the association between an exposure E and a disease D is biased if the odds ratio is systematically over- or under-estimated. In this section, three types of bias have been described: selection bias, information bias and confounding. Selection bias arises from the selection of individuals for study. Information bias results from the collection of information from the individuals selected. Confounding arises when a third variable C, known as a confounder, is associated with both E and D. Confounding can seriously bias the odds ratio and even reverse its direction, a phenomenon known as Simpson's paradox.

Exercise on Section 6

Exercise 6.1 Birth weight and hospital infection

Infections acquired in hospital are a substantial cause of mortality. A cohort study was undertaken to quantify the association between hospital-acquired infection and mortality in a neonatal intensive care unit.

Birth weight may be associated both with mortality and the acquisition of an infection. Accordingly, the data in Table 6.13 are stratified by birth weight (in grams).

Freeman, J., Goldmann, D.A. and McGowan, J.E. (1988) Methodologic issues in hospital epidemiology. IV. Risk ratios, confounding, effect modification and the analysis of multiple variables. *Reviews of Infectious Diseases*, **10**, 1118–1141.

Table 6.13 Hospital-acquired infections and mortality by birth weight

Birth weight (grams)	Hospital infection	Died	Alive	Total
< 1000	Yes	12	13	25
	No	10	20	30
1000–1499	Yes	12	30	42
	No	24	83	107
1500–1999	Yes	7	11	18
	No	18	124	142
2000+	Yes	15	38	53
	No	52	426	478

(a) Calculate the odds ratios for the association between hospital-acquired infection and mortality, for each birth weight stratum.

(b) Aggregate the data in Table 6.13 over birth weight. Calculate the odds ratio based on the aggregated table.

(c) Compare the odds ratios you obtained in parts (a) and (b). How did confounding by birth weight affect the apparent direction of association in these data?

7 Stratified analyses

In Section 6, you saw examples of confounding in which aggregation of the data reversed the direction of the association. You learned that confounding may be identified by stratifying the data by the confounding variable, and by calculating separate odds ratios for each stratum. However, an unsatisfactory aspect of this procedure is that it results in several stratum-specific odds ratios. One method of combining these estimates into a single summary measure of association is described in Subsection 7.1; this is the *Mantel–Haenszel method*. An application of this method to case-control studies is discussed in Subsection 7.2. The question of when it is appropriate to use the Mantel–Haenszel method is the subject of Subsection 7.3.

7.1 The Mantel–Haenszel odds ratio

In this subsection, you will learn how to summarize the odds ratios from several 2×2 tables. A similar procedure may be applied to relative risks, but for simplicity only odds ratios will be considered here.

Example 7.1 *Marital status, alcohol consumption and fatal car accidents*

In Example 3.3, data from a case-control study of the association between alcohol consumption and fatal car accidents in New York were introduced. Exposure was defined as a blood alcohol level of 100 mg% or greater. Cases were drivers who were killed in car accidents for which they were considered to be responsible. Controls were obtained by selecting drivers passing the locations where the accidents of the cases occurred, at the same time of day and on the same day of the week. There were 24 cases and 154 controls.

The odds ratio was 25.55, with 95% confidence interval $(8.68, 75.19)$, indicating a strong positive association between blood alcohol level and dying in a car crash.

The investigators also stratified the data by marital status, which they believed may be associated with both alcohol consumption patterns and with accident-proneness. Thus marital status may confound the association between alcohol consumption and death in car accidents. The stratified data are in Table 7.1.

Table 7.1 Alcohol consumption and fatal car accidents by marital status

(a) Married

Alcohol level	Cases	Controls	Total
$\geq 100\,\text{mg}\%$	4	5	9
$< 100\,\text{mg}\%$	5	103	108
Total	9	108	117

(b) Not married

Alcohol level	Cases	Controls	Total
$\geq 100\,\text{mg}\%$	10	3	13
$< 100\,\text{mg}\%$	5	43	48
Total	15	46	61

The stratum-specific odds ratios are as follows:

$$\widehat{OR}_{\text{Married}} = \frac{4 \times 103}{5 \times 5} = 16.48, \quad \widehat{OR}_{\text{Not married}} = \frac{10 \times 43}{3 \times 5} \simeq 28.67.$$

The odds ratio for the aggregated data is 25.55. Since this lies between the two stratum-specific odds ratios, both of which are very large and greater than 1, it is clear that confounding by marital status has not reversed the apparent direction of association. However, it may still have some effect on the magnitude of the odds ratio, so the value 25.55 may not be reliable. How can the stratum-specific odds ratios 16.48 and 28.67 be combined to produce an 'average' value that is not affected by confounding? ◆

Suppose that the stratifying variable has k levels numbered $1, 2, \ldots, k$, and that the data for the ith stratum are as shown in Table 7.2.

Table 7.2 Data from the ith stratum of a cohort study or a case-control study

Exposure category	Disease/Cases	No disease/Controls	Total
Exposed	a_i	b_i	$a_i + b_i$
Not exposed	c_i	d_i	$c_i + d_i$
Total	$a_i + c_i$	$b_i + d_i$	N_i

Then the estimated odds ratio for the ith stratum is

$$\widehat{OR}_i = \frac{a_i \times d_i}{b_i \times c_i}.$$

Suppose that the underlying odds ratios for the k strata are equal, so that the $\widehat{OR}_i$ are all estimates of some common odds ratio OR. How should the k estimates $\widehat{OR}_i$ be combined? A general approach is to use a weighted average with weights w_i that sum to 1:

$$\widehat{OR} = \sum_{i=1}^{k} w_i \widehat{OR}_i, \quad \sum_{i=1}^{k} w_i = 1.$$

For example, for the alcohol and fatal car accident data of Example 7.1, there are two strata relating to the marital status of the drivers (married or not married), so $k = 2$. Choosing $w_1 = w_2 = 0.5$, so that the two estimated odds ratios are equally weighted, gives the average odds ratio:

$$\widehat{OR} = 0.5 \times \widehat{OR}_1 + 0.5 \times \widehat{OR}_2$$
$$\simeq 0.5 \times 16.48 + 0.5 \times 28.67$$
$$\simeq 22.58.$$

Equal weights are not ideal: it would make sense, for example, to give greater weight to larger strata. Several different methods have been proposed. One popular method with good statistical properties was devised in 1959 by Nathan Mantel and William Haenszel, giving rise to what is now called the **Mantel–Haenszel odds ratio**:

$$\widehat{OR}_{MH} = \frac{\sum a_i d_i / N_i}{\sum b_i c_i / N_i}.$$

The summations are over the k strata. It is straightforward to show that the Mantel–Haenszel odds ratio is a weighted average of the stratum-specific odds ratios, with weights

$$w_i = \frac{b_i c_i / N_i}{\sum b_i c_i / N_i}.$$

The details are not important and have been omitted.

The Mantel–Haenszel odds ratio estimates the odds ratio for association between the exposure and the disease, **adjusting for** (that is, removing) the possible confounding effect of the stratifying variable.

Another popular method, which uses a different set of weights, is *logistic regression*, a regression method applicable to data in the form of proportions. This method will not be discussed in this course.

Example 7.2 The Mantel–Haenszel odds ratio for the alcohol and fatal car accident data

Table 7.1 contains data on the association between blood alcohol levels and fatal car accidents, stratified by marital status. The Mantel–Haenszel odds ratio for the association between blood alcohol level and fatal car accidents, adjusted for marital status, is given by

$$\widehat{OR}_{MH} = \frac{\sum a_i d_i / N_i}{\sum b_i c_i / N_i} = \frac{(4 \times 103)/117 + (10 \times 43)/61}{(5 \times 5)/117 + (3 \times 5)/61} \simeq \frac{10.5705}{0.4596} \simeq 23.00.$$

The value obtained by aggregating the data is 25.55. Note that since this is close to the Mantel–Haenszel adjusted odds ratio, it can reasonably be concluded that marital status is not a serious confounder. ♦

Section 7 Stratified analyses

When data are stratified, the extent of confounding can be investigated by calculating both the odds ratio based on the aggregated data, which is often called the **crude** or **unadjusted** odds ratio, and the Mantel–Haenszel odds ratio. If their values are similar, then it is unlikely that the stratifying variable is a confounder. However, if they are very different, then there is evidence of confounding.

It is generally recommended that, to correct for potential confounding, the common odds ratio is estimated using the Mantel–Haenszel estimator. However, there are some exceptions to this recommendation. In particular, if the stratifying variable C lies on the causal pathway from E to D (that is, E causes C, which in turn causes D), then the analysis should not be adjusted for the effect of variable C. This is represented schematically in Figure 7.1. (The arrows represent causal associations.) You will not encounter any such special cases in this course.

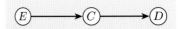

Figure 7.1 A variable C lying on the causal pathway from E to D

The calculation and interpretation of the Mantel–Haenszel odds ratio is summarized in the following box.

The Mantel–Haenszel odds ratio

Suppose that data on the association between an exposure E and a disease D have been stratified by a variable C, and that the data for stratum i from a cohort study or a case-control study are as in the table below.

Exposure category	Disease/Cases	No disease/Controls
Exposed	a_i	b_i
Not exposed	c_i	d_i

Suppose that the underlying stratum-specific odds ratio is the same for all strata. Then the common odds ratio OR is estimated by the Mantel–Haenszel odds ratio $\widehat{OR}_{MH}$, which is given by

$$\widehat{OR}_{MH} = \frac{\sum a_i d_i / N_i}{\sum b_i c_i / N_i}, \qquad (7.1)$$

where $N_i = a_i + b_i + c_i + d_i$, and the summation is over all the strata.

This estimates the strength of the association between E and D, adjusted for the possible confounding effect of C.

Activity 7.1 Adjusting for stone size

Data on the success of two treatments for kidney stones — keyhole surgery and open surgery — were discussed in Example 6.5. The data are reproduced in Table 7.3.

Table 7.3 Treatment of kidney stones by open surgery and keyhole surgery

(a) Small stones ($n = 357$)

Treatment	Success	Failure	Total
Keyhole surgery	234	36	270
Open surgery	81	6	87

(b) Large stones ($n = 343$)

Treatment	Success	Failure	Total
Keyhole surgery	55	25	80
Open surgery	192	71	263

The association between treatment method and outcome is severely confounded by stone size: the crude odds ratio for success with keyhole surgery is 1.34, whereas the stratum-specific odds ratios are 0.48 and 0.81 for small and large stones, respectively.

It is common practice to use n to denote the size of the samples, rather than N_1, N_2, which could have been used here.

(a) Obtain the Mantel–Haenszel odds ratio for the association between treatment success and keyhole surgery, adjusted for stone size.

(b) How does this estimate compare with the stratum-specific estimates and the crude odds ratio?

So far all the examples have involved two strata. The Mantel–Haenszel method applies for arbitrary numbers of strata. Activity 7.2 involves an example with four strata.

Activity 7.2 Water fluoridation and tooth decay

Water fluoridation at an appropriate concentration has long been known to protect against dental caries (tooth decay). However, artificial fluoridation of the water supply remains controversial. A study of the impact of water fluoridation on dental caries was undertaken in Scotland.

In this cohort study, two groups of children aged between 8 and 12 years from five coastal communities of Morayshire were compared. One group included 86 children from Burghead, Findhorn and Kinloss where the water supply is naturally fluoridated at 1 ppm (part per million). These constitute the exposed group. The unexposed group included 173 children from Buckie and Portessie, where the water supply is not fluoridated. The 'disease' D is dental caries. The data are in Table 7.4, stratified by age: there are four age groups.

Stephen, K.W., Macpherson, L.M.D., Gilmour, W.H., Stuart, R.A.M., Merrett, M.C.W. (2002) A blind caries and fluorosis prevalence study of schoolchildren in naturally fluoridated and nonfluoridated townships of Morayshire, Scotland. *Community Dentistry and Oral Epidemiology*, **30**, 70–79.

Table 7.4 Water fluoridation and dental caries in children aged 8–12 years

(a) Age 8 years ($n = 61$)

Water type	With caries	Without caries	Total
Fluoridated	5	25	30
Not fluoridated	8	23	31

(b) Age 9 years ($n = 67$)

Water type	With caries	Without caries	Total
Fluoridated	0	17	17
Not fluoridated	17	33	50

(c) Age 10 years ($n = 56$)

Water type	With caries	Without caries	Total
Fluoridated	5	13	18
Not fluoridated	24	14	38

(d) Age 11–12 years ($n = 75$)

Water type	With caries	Without caries	Total
Fluoridated	5	16	21
Not fluoridated	29	25	54

(a) Obtain the Mantel–Haenszel odds ratio for the association between water fluoridation and dental caries, adjusted for age.

(b) Aggregate the data and obtain the crude odds ratio.

(c) What is the apparent impact of fluoridation on dental caries? Compare the estimates you obtained in parts (a) and (b). Is there evidence of confounding?

Throughout this subsection, it has been assumed that OR, the underlying odds ratio for the association between the exposure E and disease D, is the same at all levels of the stratifying variable. This common value may be estimated by the Mantel–Haenszel odds ratio $\widehat{OR}_{MH}$. It is also possible to obtain confidence intervals for the common odds ratio OR from the stratified tables, and to test the hypothesis of no association ($OR = 1$) using a version of the chi-squared test statistic known as the Mantel–Haenszel chi-squared test statistic. However, the expressions for these are cumbersome and not particularly enlightening. So discussion of confidence intervals and tests for stratified tables will be deferred to the SPSS session in Section 9.

7.2 Matching in case-control studies

Suppose that you are involved in planning a case-control study of the impact of physical exertion on triggering heart attacks. For example, you might plan to sample both cases and controls from persons admitted to hospital, and the exposure of interest might be defined as physical exertion in the 24-hour period prior to admission. Early on in designing the study, you will need to consider what potential confounding variables you might need to take into account. For example, age and sex are potential confounders, since they are likely to be related both to the exposure (physical exertion) and to the outcome (suffering a heart attack).

How should you proceed? One approach is to collect data on these potential confounders, and stratify the analysis as described in Subsection 7.1. However, this relies to some extent on getting adequate numbers in each stratum. In a case-control study it is possible to incorporate the stratification into the study design from the outset. This procedure is known as **matching**. For each case, controls are selected so that they *match* the case with respect to the confounding variables. To match on age group and sex, for example, means that for each case, one or more controls are chosen of similar age and of the same sex as the case.

Example 7.3 Autism and MMR vaccine

In 1998 a hypothesis was formulated that measles, mumps and rubella (MMR) vaccine may in some instances be associated with the development of autism. This resulted in a long and heated controversy, and a substantial drop in MMR vaccination rates in some parts of the UK. It also resulted in numerous epidemiological studies, none of which confirmed the original hypothesis.

In one study, the possibility of an association between MMR vaccination and autism was investigated in a large case-control study. Cases included 1294 children with symptoms of pervasive developmental disorder, which includes autism, and 4469 controls, all selected from a database of general practitioner records. In order to control for potential confounding, the controls were individually matched to the cases. For each case, up to five controls were selected from the same medical practice as the case; these controls were of the same sex as the case and had a birth date within one year of that of the case.

Smeeth, L., Cook, C., Fombonne, E. *et al.* (2004) MMR vaccination and pervasive developmental disorders: a case-control study. *Lancet*, **364**, 963–969.

Each case and matched control was classified as having received MMR vaccination or not having received MMR vaccination before the date of diagnosis of the case. The odds ratio for the association between pervasive developmental disorder and MMR vaccination was 0.86, with 95% confidence interval (0.68, 1.09). Since cases and controls were matched on birth year (and hence age group), sex and medical practice, these variables are controlled for automatically by the study design. ♦

Matching can be considered as a (somewhat extreme) form of stratification, in which each stratum comprises a **matched case-control set**, namely a case and the controls that have been matched to it. Thus there are as many strata as there are cases. The standard method of analysing matched case-control studies is by stratifying the analysis by the matched sets. The Mantel–Haenszel odds ratio may be used to obtain a summary odds ratio adjusted for the matching variables.

The rest of this subsection concerns the special case of a **1–1 matched case-control study**; this is a case-control study in which a single control is matched to each case. You will learn how to estimate the Mantel–Haenszel odds ratio for 1–1 matched case-control studies, calculate confidence intervals for the odds ratio and test for no association using a test called McNemar's test.

Suppose that data on k case-control pairs have been collected. So there are k cases and k individually matched controls. These pairs can be arranged as in Table 7.5.

Table 7.5 Standard layout for a 1–1 matched case-control study

		Controls	
		Exposed	Not exposed
Cases	Exposed	e	f
	Not exposed	g	h

The entries in the table refer to numbers of pairs. Thus, there are e pairs in which both the case and the control are exposed. The total number of pairs is k, so $e + f + g + h = k$. Note that this layout differs from the standard layout for a case-control study.

Alternatively, though less succinctly, the data can be regarded as consisting of k strata numbered $1, 2, \ldots, k$, each involving one case and one control, and may be set out as in Table 7.6.

Table 7.6 Alternative layout for a 1–1 matched case-control study

(1) Matched set 1

	Case	Control
Exposed	a_1	b_1
Not exposed	c_1	d_1
Total	1	1

$\cdots$

(k) Matched set k

	Case	Control
Exposed	a_k	b_k
Not exposed	c_k	d_k
Total	1	1

So, using Formula (7.1), the Mantel–Haenszel odds ratio is given by

$$\widehat{OR}_{MH} = \frac{(a_1 d_1/2) + \cdots + (a_k d_k/2)}{(b_1 c_1/2) + \cdots + (b_k c_k/2)}$$

$$= \frac{a_1 d_1 + \cdots + a_k d_k}{b_1 c_1 + \cdots + b_k c_k}. \tag{7.2}$$

Consider the numerator of this expression. Each of the quantities a_i and d_i is equal to either 1 or 0 because each matched set includes only one case and one control. The term $a_1 d_1$ is equal to 1 when $a_1 = d_1 = 1$; otherwise, it is zero. Now $a_1 = d_1 = 1$ when the case in the first matched set is exposed and the control is not exposed. So the term $a_1 d_1$ is equal to 1 only when the case is exposed and the control is not exposed. Similarly, for each i, the term $a_i d_i$ is equal to 1 only when the case in the ith matched set is exposed and the control is not exposed. So the numerator in the expression (7.2) is equal to the number of pairs with case exposed and control not exposed; this is denoted f in Table 7.5. So

$$a_1 d_1 + \cdots + a_k d_k = f.$$

Similarly, if g is as defined in Table 7.5, then

$$b_1 c_1 + \cdots + b_k c_k = g.$$

Hence, for a 1–1 matched case-control study, the expression for the Mantel–Haenszel odds ratio reduces to

$$\widehat{OR}_{MH} = \frac{f}{g}. \tag{7.3}$$

Case-control pairs in which the case is exposed and the control is not exposed, or vice versa, are called **discordant pairs**. Thus the Mantel–Haenszel odds ratio is equal to the ratio of the numbers of the two types of discordant pairs.

Example 7.4 Blood clots and contraceptive pill use

Widespread use of oral contraceptives ('the Pill') in the 1960s soon led to concerns that using them could increase the risk of thromboembolism (blood clots). These can cause severe pain, heart attacks and stroke. A case-control study was undertaken in the United States to study the association between use of oral contraceptives and thromboembolism.

Sartwell, P.E., Masi, A.T., Arthes, F.G., Greene, G.R. and Smith, H.E. (1969) Thromboembolism and oral contraceptives: an epidemiologic case-control study. *American Journal of Epidemiology*, **90**, 365–380.

The study was a 1–1 matched case-control study, with 175 case-control pairs. Cases were chosen among married women aged 15–44 who had been discharged alive from hospital for a first occurrence of thromboembolism. Controls were selected from married women discharged alive from the same hospital during the same period, for conditions not thought to be related to contraceptive pill use. For each case, a control was selected of the same age, place of residence, ethnicity, number of prior pregnancies, and type of health insurance.

The exposure is oral contraceptive pill use within one month of admission to hospital. The data are in Table 7.7.

Table 7.7 Contraceptive pill use and hospital admission for blood clots

		Controls	
		Exposed	Not exposed
Cases	Exposed	10	57
	Not exposed	13	95

The Mantel–Haenszel odds ratio, adjusted for the matching factors, is given by Formula (7.3):

$$\widehat{OR}_{MH} = \frac{f}{g} = \frac{57}{13} \simeq 4.38.$$

This indicates a rather strong positive association between contraceptive pill use in the 1960s and thromboembolism. ♦

The composition of contraceptive pills has changed substantially since the 1960s.

Activity 7.3 Heart attacks among Navajo Indians

Navajo Indians have in the past suffered lower rates of heart disease than non-Native Americans. A case-control study was undertaken of risk factors for heart attacks among Navajo Indians. The cases were 144 persons discharged from hospital with a diagnosis of acute myocardial infarction (heart attack). Each case was individually matched on age and sex with a person discharged from the same hospital with a diagnosis of gall-bladder disease, a relatively common complaint among Navajo Indians.

Coulehan, J.L., Lerner, G., Helzlsouer, K., Welty, T.K. and McLaughlin, J. (1986) Acute myocardial infarction among Navajo Indians, 1976–83. *American Journal of Public Health*, **76**, 412–414.

Several exposures were considered in this study, including a prior diagnosis of diabetes. For 9 of the 144 case-control pairs, both case and control suffered from diabetes, while for 82 pairs neither had diabetes. In 37 of the 53 discordant pairs, only the case suffered from diabetes, while in the remaining 16, only the control had diabetes.

(a) Draw up a table showing the distribution of pairs according to exposure status of case and control.

(b) Estimate the odds ratio for the association between diabetes and heart attacks in Navajo Indians, adjusting for age, sex and hospital.

An interesting feature of the Mantel–Haenszel odds ratio for matched pairs is that it depends only on f and g, the numbers of discordant pairs. The values of e and h are irrelevant, in the sense that they contain no information about the association. Similarly, it can be shown that the standard error of the logarithm of the Mantel–Haenszel odds ratio for matched pairs also depends only on f and g. Provided that f and g are sufficiently large, the standard error of $\log(\widehat{OR}_{MH})$ may be estimated by

$$\hat{\sigma} = \sqrt{\frac{1}{f} + \frac{1}{g}}. \tag{7.4}$$

Thus, using an argument similar to that used in Subsection 2.2, approximate $100(1-\alpha)\%$ confidence limits for the odds ratio are given by

$$OR^- = \widehat{OR}_{MH} \times \exp\left(-z\sqrt{\frac{1}{f} + \frac{1}{g}}\right),$$

$$OR^+ = \widehat{OR}_{MH} \times \exp\left(z\sqrt{\frac{1}{f} + \frac{1}{g}}\right),$$

where z is the $(1-\alpha/2)$-quantile of the standard normal distribution.

Example 7.5 Calculating confidence intervals

For the data on oral contraceptive use and blood clots of Example 7.4, $f = 57$ and $g = 13$. The estimated odds ratio is $f/g = 57/13 \simeq 4.3846$. For these data, the estimated standard error is

See Table 7.7.

$$\hat{\sigma} = \sqrt{\frac{1}{f} + \frac{1}{g}} = \sqrt{\frac{1}{57} + \frac{1}{13}} \simeq 0.3074.$$

For a 95% confidence interval, $z = 1.96$. So the 95% confidence limits are as follows:

$$OR^- = \widehat{OR}_{MH} \times \exp\left(-z\sqrt{\frac{1}{f} + \frac{1}{g}}\right)$$

$$\simeq 4.3846 \times \exp(-1.96 \times 0.3074)$$

$$\simeq 2.40,$$

$$OR^+ = \widehat{OR}_{MH} \times \exp\left(z\sqrt{\frac{1}{f} + \frac{1}{g}}\right)$$

$$\simeq 4.3846 \times \exp(1.96 \times 0.3074)$$

$$\simeq 8.01.$$

Thus the 95% confidence interval is $(2.40, 8.01)$. ◆

The fact that the Mantel–Haenszel odds ratio for a 1–1 matched case-control study depends only on f and g suggests a method for testing the null hypothesis $OR = 1$, known as **McNemar's test**.

Under the null hypothesis $OR = 1$, a discordant pair is as likely to be one in which the case is exposed as one in which the control is exposed. Thus, among the $f + g$ discordant pairs, the event 'case exposed' may be regarded as a Bernoulli trial with probability $\frac{1}{2}$ and the number of such pairs out of the $f + g$ observed has the binomial distribution $B(f + g, \frac{1}{2})$. The upper and lower tails of this distribution, corresponding to values of f at least as extreme as that observed (and hence producing odds ratios at least as far from 1), determine the p value for the test.

The tails of the distribution are shown in Figure 7.2 for $f + g = 20$ and $f = 15$. This corresponds to an odds ratio of $15/5 = 3$.

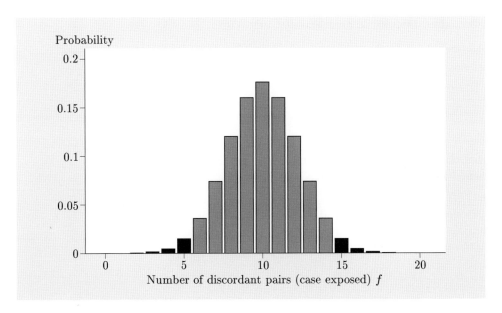

Figure 7.2 Calculating the p value for McNemar's test

Values of f between 15 and 20 give odds ratios of 3 or more, which are at least as extreme as the observed odds ratio of 3. Under the null hypothesis of no association, the distribution of f is $B(20, \frac{1}{2})$, so the upper tail probability is $P(X \geq f) = P(X \geq 15) \simeq 0.0207$. For a two-sided test, this probability is doubled to give the p value: $p \simeq 2 \times 0.0207 \simeq 0.041$. This indicates that there is moderate evidence against the null hypothesis of no association.

Calculating the binomial probabilities required to obtain the exact p value is best done using a computer. However, for all but the smallest samples, a test statistic known as **McNemar's test statistic** may be used instead of f. It is denoted χ^2 and is given by

$$\chi^2 = \frac{(|f - g| - 1)^2}{f + g}. \tag{7.5}$$

The calculation of exact p values is deferred to the SPSS session in Section 9.

The vertical bars denote the absolute value. So, for example, $|-5| = 5$.

This is the test statistic for McNemar's test for no association. Under the null hypothesis of no association ($OR = 1$), the distribution of this test statistic may be approximated by $\chi^2(1)$, the chi-squared distribution on 1 degree of freedom, provided that both f and g are sufficiently large. The interpretation of p values from McNemar's test is the same as for the chi-squared test for no association. (See Table 4.7.) The test is illustrated in Example 7.6.

In fact, it can be shown that McNemar's test statistic is a modified version of the test statistic for the chi-squared test for no association that was described in Section 4.

Example 7.6 Blood clots and oral contraceptive use

The data on oral contraceptive use and blood clots, which were described in Example 7.4, are in Table 7.7. For these data, $f = 57$ and $g = 13$. So, using (7.5), the observed value of McNemar's test statistic is

$$\chi^2 = \frac{(|57 - 13| - 1)^2}{57 + 13} \simeq 26.41.$$

The 0.999-quantile of $\chi^2(1)$ is 10.83. Since $26.41 > 10.83$, the significance probability is $p < 0.001$. Thus there is strong evidence of an association between use of oral contraceptives and blood clots.

See the table of quantiles of chi-squared distributions given in the Handbook.

Since the odds ratio is 4.38, which is greater than 1, the data from the 1969 study suggest that oral contraceptive use is strongly positively associated with the formation of serious blood clots. ♦

Activity 7.4 Heart attacks among Navajo Indians

In Activity 7.3, a 1–1 matched case-control study of diabetes and heart attacks among Navajo Indians was described. The data are summarized in Table 7.8.

Table 7.8 Diabetes and heart attacks among Navajo Indians

		Controls	
		Exposed	Not exposed
Cases	Exposed	9	37
	Not exposed	16	82

(a) In Activity 7.3, you found that the Mantel–Haenszel odds ratio for the association between diabetes and heart attacks is $2.3125 \simeq 2.31$. Obtain a 95% confidence interval for the odds ratio.

(b) Use McNemar's test to test for no association between diabetes and heart attacks among Navajo Indians.

The analysis of 1–1 matched case-control studies is summarized in the following box.

Analysis of 1–1 matched case-control studies

For a 1–1 matched case-control study, the Mantel–Haenszel estimate of the odds ratio is equal to the ratio of discordant case-control pairs. That is,

$$\widehat{OR}_{MH} = \frac{f}{g},$$

where f is the number of pairs in which only the case is exposed and g is the number of pairs in which only the control is exposed.

Approximate $100(1-\alpha)\%$ confidence limits for the odds ratio are given by

$$OR^- = \widehat{OR}_{MH} \times \exp\left(-z\sqrt{\frac{1}{f} + \frac{1}{g}}\right),$$

$$OR^+ = \widehat{OR}_{MH} \times \exp\left(z\sqrt{\frac{1}{f} + \frac{1}{g}}\right),$$

where z is the $(1-\alpha/2)$-quantile of the standard normal distribution.

McNemar's test for no association is based on the test statistic

$$\chi^2 = \frac{(|f-g|-1)^2}{f+g}.$$

Under the null hypothesis of no association, the distribution of this test statistic is approximately $\chi^2(1)$.

7.3 Interactions

The Mantel–Haenszel method for combining odds ratios from different strata, with the purpose of adjusting for a potential confounding variable C, requires that the underlying odds ratios in each stratum are the same. But what if this requirement is violated, so that, for example, the odds ratio for $C = 0$ is different from the odds ratio for $C = 1$? In this case, it is said that there is an **interaction** between the stratifying variable C and the association between the exposure E and the disease D. The concept of interaction is explained in Example 7.7 and the discussion that follows.

Example 7.7 Fenoterol, asthma severity and asthma deaths in New Zealand

An increase in deaths due to asthma was observed in New Zealand in the late 1970s, following the prescription for home use of new types of drugs known as beta-2 agonists.

A case-control study was undertaken to investigate the possible association between prescription of one particular type of medication, namely fenoterol delivered by a nebulizer, and asthma deaths. Both cases and controls were chosen among persons who were admitted to hospital for asthma. The cases comprised 117 persons with asthma who died of asthma; the controls were 468 persons with asthma who did not die of asthma. The data are in Table 7.9.

Crane, J., Pearce, N., Flatt, A. et al. (1989) Prescribed fenoterol and death from asthma in New Zealand, 1981–83: case-control study. *Lancet*, **1**, 917–922.

The estimated odds ratio is

$$\widehat{OR} = \frac{60 \times 279}{189 \times 57} \simeq 1.55.$$

This is suggestive of a moderate positive association between fenoterol prescription and asthma deaths in persons admitted to hospital for asthma.

However, one concern was that cases and controls may have differed according to the underlying severity of their asthma. Indeed, disease severity may also be associated with fenoterol prescription, and hence is a potential confounder. Accordingly, the data were stratified by variables associated with severity. One such indicator of severity is prescription for oral steroids in the previous year. The data, stratified by this variable, are shown in Table 7.10.

Table 7.9 Fenoterol and asthma deaths

Fenoterol prescribed	Cases	Controls
Yes	60	189
No	57	279
Total	117	468

Table 7.10 Fenoterol and asthma deaths, stratified by prescription for steroids

(1) No steroids prescribed

Fenoterol prescribed	Cases	Controls
Yes	34	151
No	50	213
Total	84	364

(2) Steroids prescribed

Fenoterol prescribed	Cases	Controls
Yes	26	38
No	7	66
Total	33	104

The estimated odds ratio for patients to whom oral steroids were not previously prescribed is

$$\widehat{OR}_1 = \frac{34 \times 213}{151 \times 50} \simeq 0.96.$$

In contrast, for patients to whom oral steroids were previously prescribed, the estimated odds ratio is

$$\widehat{OR}_2 = \frac{26 \times 66}{38 \times 7} \simeq 6.45.$$

Thus, in this case, the estimated odds ratios differ very substantially between strata. If this reflects a real underlying difference, it may not be sensible to combine the odds ratios into a single estimate using the Mantel–Haenszel method, since the data suggest that the effect of fenoterol differs between the two strata. ◆

In Example 7.7, there is a possible **interaction** between underlying asthma severity (as indicated by prescription of oral steroids) and the association between fenoterol use and asthma deaths. Thus, in less severe cases (for whom oral steroids were not prescribed), the odds ratio is 0.96, which is not suggestive of an association between fenoterol and asthma death. However, for people with more severe disease (who were previously prescribed oral steroids) the odds ratio is 6.45, suggesting a strong positive association.

From the statistical point of view, the term interaction is descriptive and implies nothing about causality. In this sense it is like the term 'association'. However, when a causal effect is inferred, the interacting variable is sometimes called an **effect modifier**. In the asthma example, the causal effect of fenoterol on asthma mortality is *modified* by the underlying disease severity.

When can an interaction be said to exist? Clearly, owing to random fluctuations, the estimates of OR from different strata will not generally be exactly the same even if the underlying odds ratio OR is the same across strata. Thus a statistical test is required to distinguish interaction from random variation.

A significance **test of homogeneity** is used to test the null hypothesis that $OR_1 = OR_2 = \ldots = OR_k$, where k is the number of strata and OR_i is the odds ratio for stratum i. One test with good properties is **Tarone's test for homogeneity**. SPSS will be used to carry out the test in Section 9. Under the null hypothesis of homogeneity, the distribution of Tarone's test statistic is approximately chi-squared on $k-1$ degrees of freedom. The larger the value of the test statistic (and hence the smaller the p value), the stronger is the evidence against the null hypothesis that the odds ratios are the same, and hence the stronger is the evidence that the odds ratios differ between strata.

Activity 7.5 Interpreting Tarone's test for homogeneity

Tarone's test for homogeneity of the underlying odds ratio was applied to the asthma data of Example 7.7. The value of the test statistic was 13.77. What do you conclude?

Summary of Section 7

In this section, methods for controlling confounding by stratifying an analysis have been discussed. The Mantel–Haenszel odds ratio, which is a weighted average of the stratum-specific odds ratios, has been described.

Matching is a particular form of stratification that may be applied to the design of a case-control study. You have learned how to calculate the Mantel–Haenszel estimate of the odds ratio for 1–1 matched case-control studies, and to calculate confidence intervals for the odds ratio; and you have used McNemar's test to test for no association.

An interaction is said to exist if the odds ratios in different strata differ. Tarone's test for homogeneity, which is used to test for the presence of an interaction, has been discussed briefly.

Exercises on Section 7

Exercise 7.1 Birth weight and hospital infection

In Exercise 6.1, you investigated data from a neonatal intensive care unit on hospital-acquired infections and mortality, stratified by birth weight. The data from Table 6.13 are reproduced in Table 7.11.

Table 7.11 Hospital-acquired infections and mortality by birth weight

Birth weight (grams)	Hospital infection	Died	Alive	Total
< 1000	Yes	12	13	25
	No	10	20	30
1000–1499	Yes	12	30	42
	No	24	83	107
1500–1999	Yes	7	11	18
	No	18	124	142
2000+	Yes	15	38	53
	No	52	426	478

(a) The value of the test statistic for Tarone's test for homogeneity applied to the data in Table 7.11 is 4.06. Test the null hypothesis that the odds ratios are the same across the four strata.

(b) Obtain the Mantel–Haenszel odds ratio for the association between acquisition of an infection while in hospital and death in a neonatal intensive care unit, adjusted for the effect of birth weight.

(c) In Exercise 6.1, you found that the odds ratio from the aggregated data (that is, aggregated over birth weight) is 3.14. Compare this value with the value you found in part (b). What do you conclude about the effect of confounding by birth weight on the association between hospital-acquired infection and mortality?

Exercise 7.2 Infection and stroke

It has been suggested that infection with the bacterium *Chlamydia pneumoniae* is associated with the occurrence of stroke. A 1–1 matched case-control study was conducted to investigate this hypothesis. In the study, 134 stroke patients were matched with 134 controls on age, gender and underlying medical condition. The exposure was evidence of recent infection by *Chlamydia pneumoniae*, as assessed by tests on blood samples.

Tanne, D., Haim, M., Boyko, V. et al. (2003) Prospective study of *Chlamydia pneumoniae* IgG and IgA seropositivity and risk of incident ischemic stroke. *Cerebrovascular Diseases*, **16**, 166–170.

The data from the study are shown in Table 7.12.

Table 7.12 Stroke and recent infection

		Controls Exposed	Controls Not exposed
Cases	Exposed	88	21
Cases	Not exposed	16	9

(a) Estimate the Mantel–Haenszel odds ratio for the association between recent infection and stroke.

(b) Calculate an approximate 95% confidence interval for the odds ratio.

(c) Test the null hypothesis of no association between recent infection and stroke using McNemar's test.

(d) Summarize your conclusions.

Exercise 7.3 Arsenic exposure and lung cancer in copper-smelter workers

In a case-control study of the impact of exposure to arsenic on lung cancer among Swedish copper-smelter workers, data were collected on 107 male workers who died of cancer and 214 male workers who died of other causes. Lifetime exposure to arsenic was estimated (in mg/m^3yr) using company records. An individual was classified as exposed if he was exposed to 15 or more mg/m^3yr.

Järup, L. and Pershagen, G. (1991) Arsenic exposure, smoking, and lung cancer in smelter workers – a case-control study. *American Journal of Epidemiology*, **134**, 545–551.

Since smoking is a known risk factor for lung cancer, the data were stratified according to whether or not cases and controls smoked. The data are in Table 7.13.

Table 7.13 Arsenic exposure and death from lung cancer in smokers and non-smokers

	Arsenic exposure (mg/m^3yr)	Cases	Controls
Non-smokers	Exposed (≥ 15)	8	28
	Not exposed (< 15)	3	46
Smokers	Exposed (≥ 15)	40	47
	Not exposed (< 15)	56	93

(a) Estimate the stratum-specific odds ratios for non-smokers and smokers. In your view, is it sensible to calculate the Mantel–Haenszel odds ratio for non-smokers and smokers combined?

(b) The value of the test statistic for Tarone's test for homogeneity is 2.27. Test the null hypothesis that the underlying odds ratio for association between arsenic exposure and lung cancer is the same for smokers and non-smokers.

(c) Contrast your results from parts (a) and (b). What do you conclude?

8 From association to causation

The purpose of most epidemiological investigations is to throw light on *causes* of disease and ill health. However, the statistical methods discussed so far can only help to establish whether an exposure is *associated* with a disease. And, as has often been repeated, association does not imply causation.

So how should we proceed? The first step is to examine possible sources of bias and confounding, as described in Sections 6 and 7, and their likely impact on the results. Such an examination exhibits the limitations of a study and helps to qualify the strength of the statistical evidence.

Once serious bias has been ruled out, positive evidence for a causal link is examined, from other studies and from other scientific disciplines. Some of this evidence may be statistical, though usually other types of evidence are also considered, for example from biology. In Subsection 8.1, some criteria for establishing causation are set out. In Subsection 8.2, statistical methods are described relating to one of these criteria.

8.1 Bradford Hill's criteria for causation

In 1965, the statistician Austin Bradford Hill (see Figure 8.1) proposed a list of criteria to help in assessing whether an association between an exposure and a disease is likely to be causal. (He actually described the criteria more loosely as viewpoints.) However, fulfilment of these criteria does not demonstrate a causal link.

Hill, A.B. (1965) The environment and disease: Association or causation? *Proceedings of the Royal Society of Medicine*, **58**, 295–300.

Three of these criteria can be assessed using statistical analysis. They are as follows.

- ◇ **Strength of association**
  A strong association (that is, an odds ratio well above or well below 1) is more likely to be causal than a weak association. This is because a weak association is more likely to be the result of bias, for example through confounding due to some variable that has not been taken into account in the analysis.

- ◇ **Dose-response relationship**
  Evidence for causality is stronger if the risk of disease increases as the degree of exposure increases. For example, the fact that people who smoke many cigarettes per day have a higher risk of lung cancer than people who smoke only a few cigarettes a day provides further evidence that the association between smoking and lung cancer is causal.

- ◇ **Consistency of the association**
  Evidence for a causal association is stronger if the association is present in different studies, undertaken by different people, in different places and under different circumstances. Conversely, a single study is unlikely to be sufficient to infer causality.

Figure 8.1 Sir Austin Bradford Hill (1897–1991)
Reproduced by permission of the Royal Statistical Society

The other criteria include, for example, temporality (exposure must precede disease), biological plausibility and lack of conflict with wider scientific understanding.

Of the three criteria listed above, the first may be addressed by estimating the odds ratio for the association between exposure and disease, and calculating a confidence interval. In Subsection 8.2, the statistical analysis of dose-response data will be described. The third criterion will be considered in Part III.

8.2 Dose-response analysis

In some cases the exposure E can be quantified. For example, if the exposure is smoking, this can be quantified by the average number of cigarettes smoked per day. If the exposure is consumption of alcohol, this can be quantified by the number of units of alcohol consumed per week. A quantified exposure is called a **dose**: for example, smoking is an exposure, whereas 20 cigarettes per day is a dose of that exposure.

A **dose-response relationship** is said to exist between an exposure E and a disease D if the risk (or odds) of disease varies according to the dose of that exposure. This can be represented by a **dose-response curve**, such as that in Figure 8.2. The curve shown in Figure 8.2 represents an increasing dose-response relationship: the larger the dose, the larger the risk. In studies such as cohort studies and case-control studies, evidence of a dose-response relationship can be obtained by subdividing the exposed individuals according to exposure dose. Then the odds ratio can be calculated for each dose, relative to individuals in the unexposed group (or, in some cases, the group with lowest exposure dose). These are **dose-specific** odds ratios. This is illustrated in Example 8.1.

Some exposures cannot be quantified. For example, gender is a risk factor for death (women live longer than men on average), but gender is either male or female and cannot be quantified.

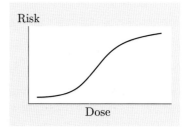

Figure 8.2 A dose-response curve

Example 8.1 Smoking and lung cancer

In Example 3.1, the 1950 case-control study by Doll and Hill of smoking and lung cancer was described. The study included 649 men with lung cancer (the cases) and 649 men without lung cancer (the controls). The overall odds ratio was 14.04, with 95% confidence interval $(3.33, 59.31)$.

The average daily dose of tobacco smoked immediately before the time of the illness was calculated. The units used were cigarettes per day; other types of tobacco were converted to 'cigarette equivalents'. The data for males are presented in Table 8.1 with the exposure classified according to dose. As only two cases did not smoke, non-smokers (0 cigarettes per day) were grouped with persons smoking 1–4 cigarettes per day.

Table 8.1 Smoking and lung cancer, by dose

Average number of cigarettes per day	Cases	Controls
50+	32	13
25–49	136	71
15–24	196	190
5–14	250	293
0–4	35	82
Total	649	649

To investigate the effect of exposure dose on the strength of association, dose-specific odds ratios are calculated relative to the lowest exposure level. For example, for the highest dose group, namely individuals who smoked 50 or more cigarettes per day, the odds ratio is calculated using the row corresponding to the 50+ exposure group, and the row corresponding to the 0–4 group. Thus

$$\widehat{OR}_{50+} = \frac{32 \times 82}{13 \times 35} \simeq 5.77.$$

Similarly, the dose-specific odds ratio for the 25–49 cigarettes group is

$$\widehat{OR}_{25-49} = \frac{136 \times 82}{71 \times 35} \simeq 4.49.$$

Repeating this for each dose gives the dose-specific odds ratios in Table 8.2 (overleaf).

Table 8.2 Smoking and lung cancer: dose-specific odds ratios

Average number of cigarettes per day	Cases	Controls	Odds ratio
50+	32	13	5.77
25–49	136	71	4.49
15–24	196	190	2.42
5–14	250	293	2.00
0–4	35	82	1.00
Total	649	649	

These odds ratios are all less than 14.04, the value of the overall odds ratio for smoking compared to non-smoking. This is because non-smokers have been combined with light smokers.

The dose-specific odds ratios (relative to the 0–4 category) display a clear trend, strongly suggestive of an increasing dose-response relationship: the more cigarettes smoked, the stronger the association with lung cancer, and hence the higher the risk of lung cancer. In turn, this supports (but does not prove) the hypothesis that the association between smoking and lung cancer is causal. ♦

In Example 8.1, the dose could be quantified (on a scale measured by numbers of cigarettes smoked per day). In other instances, it may be more difficult to quantify the exposure dose. However, it might be possible to group the exposures in ordered categories — for example, No exposure, Low exposure, Medium exposure, High exposure. When this is the case, the dose-response relationship can be investigated in exactly the same way as if the dose were quantified numerically. Activity 8.1 provides an example of such data.

Activity 8.1 Post-traumatic stress disorder among US veterans

In Example 2.1, data were presented from a cohort study conducted among United States army veterans to investigate the association between post-traumatic stress disorder (PTSD) and deployment to the Persian Gulf during the 1991 Gulf War. The exposed group included 6617 veterans who were deployed to the Persian Gulf. The unexposed group included 2963 veterans who were deployed to areas other than the Persian Gulf.

In fact, these data form part of a more detailed data set of 12 424 veterans, grouped in six categories according to the level of stress they experienced during the 1991 Gulf War. The full data are in Table 8.3. The stress levels in Table 8.3 are labelled so as to indicate an ordering, from lowest (Minimal) to highest (Extreme). However, the stress levels are not quantified.

Table 8.3 Post-traumatic stress disorder among US veterans by level of stress experienced

Stress level	PTSD	No PTSD	Total
Extreme	174	595	769
Very severe	260	1155	1415
Severe	362	2688	3050
High	97	1286	1383
Moderate	180	2783	2963
Minimal	95	2749	2844

(a) Calculate the odds ratios for the association between stress level and PTSD, for each of the first five stress levels relative to the Minimal category.

(b) What do you observe? What might you conclude about the presence or otherwise of an increasing dose-response relationship between stress and risk of PTSD?

Section 8 From association to causation

In Example 8.1 and Activity 8.1, you observed that the odds ratio increased with dose or exposure level. But could these patterns have arisen by chance? To help decide the issue a statistical test is required. How can such a test be constructed?

Consider the data of Table 8.2. One approach might be to use Tarone's test for homogeneity, described in Subsection 7.3. Such an approach, however, would not be valid: since all the odds ratios are calculated relative to the same lowest exposure category, the odds ratios are not independent. However, the *odds* of disease for each exposure category *are* independent.

The odds of disease for the data in Table 8.2 are shown in Table 8.4.

Table 8.4 Smoking and lung cancer: dose-specific odds ratios

Average number of cigarettes per day	Cases	Controls	Odds of disease	log(odds)
50+	32	13	2.462	0.901
25–49	136	71	1.915	0.650
15–24	196	190	1.032	0.031
5–14	250	293	0.853	−0.159
0–4	35	82	0.427	−0.851
Total	649	649		

For example, the estimated odds of disease for those who smoke 0–4 cigarettes per day are

$$\widehat{OD}_{0-4} = 35/82 \simeq 0.4268.$$

In contrast, the estimated odds for those who smoke 50+ cigarettes per day are

$$\widehat{OD}_{50+} = 32/13 \simeq 2.4615.$$

This is 5.77 times higher.

Note that interest lies in *comparing* the odds of disease at different doses, not in their absolute values. Such comparisons are valid for both cohort studies and case-control studies.

A reasonable starting point when testing for a dose-response relationship is to plot the odds against the doses. In fact, it is more convenient to plot the logarithms of the odds. These are expressed on a scale from $-\infty$ to $+\infty$, whereas the odds are on a scale from 0 to ∞. Figure 8.3 shows such a plot, using the average doses 2, 9.5, 19.5, 37 and (rather arbitrarily) 50 on the horizontal axis. The plot indicates that there exists a roughly linear relationship between the log(odds) and the dose.

Also shown in Figure 8.3 is the regression line

$$\log(OD_i) = \alpha + \beta x_i,$$

where OD_i is the odds of disease at dose x_i. A positive slope corresponds to an increasing dose-response relationship (and a negative slope corresponds to a decreasing dose-response relationship). And the steeper the regression line, the stronger is the dose-response relationship. If the slope of the regression line were close to zero, it would be reasonable to conclude that there was little evidence of a linear dose-response relationship. Thus a significance test for the presence of a linear dose-response relationship could be based on the null hypothesis of zero trend, namely $\beta = 0$, where β is the slope of the regression line.

The **chi-squared test for no linear trend** is based on a test of the slope of a regression line. Under the null hypothesis of zero trend, that is $\beta = 0$, for large samples the distribution of the test statistic is approximately chi-squared on 1 degree of freedom. The larger the value of the test statistic (and hence the smaller the p value), the stronger is the evidence for a linear dose-response relationship. The details will be omitted here. You will learn how to carry out the test using SPSS in Chapter 5 of *Computer Book 1*.

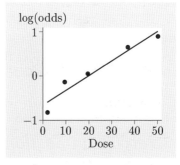

Figure 8.3 A scatterplot of the log(odds) for lung cancer against number of cigarettes smoked, and the regression line

Linear regression is a statistical technique for fitting a straight line through a set of points. The regression line is the 'best fitting' line through the points. You do not need to know how to calculate its equation in this course.

In fact, there are several different tests for no linear trend, all of which give similar results in large samples.

This linear trend test can also be used when the doses have not been measured precisely, provided that the exposure categories can be ordered according to the dose received. In this case, the doses x_i are simply replaced by ranks.

Activity 8.2 Dose-response curve for PTSD and stress

(a) The stress levels in Table 8.3 are ordered. Label them $i = 1, 2, 3, 4, 5, 6$ with 1 corresponding to Minimal and 6 to Extreme. Calculate the logarithms of the estimated odds, $\log(\widehat{OD}_i)$, and plot them against i. What do you observe?

(b) The chi-squared test for no linear trend gives the value 455.15 for the test statistic. What do you conclude?

Evidence of a linear dose-response relationship supports a causal association. However, it should be emphasized that such evidence does not constitute proof of causality. Conversely, causal associations can exhibit non-linear dose-response relationships. For example, moderate consumption of red wine is believed to be beneficial, whereas excessive consumption is detrimental to health.

Summary of Section 8

In this section, Bradford Hill's criteria for establishing causation have been described. One of these criteria relates to the presence of a dose-response relationship. You have seen that a scatterplot is a good starting point when testing for a dose-response relationship. A test for no linear trend has been discussed briefly.

Exercise on Section 8

Exercise 8.1 Dose-response analysis of arsenic and lung cancer data

In Exercise 7.3, a case-control study of the impact of exposure to arsenic on lung cancer among Swedish copper-smelter workers was described. Exposures to arsenic were quantified and grouped into four categories: under $0.25 \text{ mg/m}^3\text{yr}$, 0.25 to less than 15 ($0.25\text{–}15^-$), 15 to less than 100 ($15\text{–}100^-$), and 100 or more. The numbers of cases and controls in each category (smokers and non-smokers combined) are in Table 8.5.

Table 8.5 Dose of arsenic and death from lung cancer

Dose ($\text{mg/m}^3\text{yr}$)	Cases	Controls
≥ 100	12	9
$15\text{–}100^-$	36	66
$0.25\text{–}15^-$	45	104
< 0.25	14	35

(a) Calculate the dose-specific odds ratio for each of the first three dose levels relative to the lowest dose. What do you observe?

(b) The chi-squared test for no linear trend gives the value 2.97 for the test statistic. Test the null hypothesis that there is no linear dose-response relationship between arsenic dose and death from lung cancer.

9 Analysis of stratified tables in SPSS

In this section, you will learn how to use SPSS to do calculations of the type described in Sections 7 and 8.

Refer to Chapters 3, 4 and 5 of Computer Book 1 for the work in this section.

Summary of Section 9

In this section, you have learned how to input stratified data in SPSS. The calculation of the Mantel–Haenszel odds ratio in SPSS has been described, together with confidence intervals and the Mantel–Haenszel test for no association. You have seen that the output produced by SPSS is extensive and includes the results of Tarone's test for homogeneity. For 1–1 matched case-control studies, you have learned how to use SPSS to calculate the exact two-sided p value for McNemar's test. Finally, you have learned how to use SPSS to test for no linear trend in dose-response.

Part III Randomized controlled trials and the medical literature

Introduction to Part III

In the examples you encountered in Parts I and II, the exposure variables included a wide range of factors that might have either a detrimental or a beneficial impact on health. These included smoking, the wearing of seat belts, being made redundant, babies' sleeping position, use of hormone-replacement therapy, arsenic exposure, political activity, alcohol consumption, and so on. A feature common to all these studies is that the investigators had no control over who was exposed and who was not. For example, in their cohort study of cannabis use and depression (described in Example 2.3), the investigators had no influence on who did or did not smoke cannabis. The individuals in the study would have had the same exposures and outcomes whether or not the study had been done.

In Sections 10 and 11, you will learn about a different type of study, in which the investigators have more control over who is exposed and who is not. These studies are known as **randomized controlled trials**, or RCTs for short. RCTs may be used to evaluate a wide range of interventions, in medicine and beyond — for example, a health education programme, a new surgical procedure, or the efficacy of a pharmaceutical drug. In Section 10, some of the distinguishing features of RCTs are described. You will learn how to calculate the sample size required for an RCT in Section 11.

An important role for the modern statistician is to assess the statistical evidence available from published studies critically. Medical and epidemiological information about the association between an exposure and disease, or about the performance of a type of drug, is often available from several different studies, which might include cohort studies, case-control studies and RCTs. The important topic of systematic reviews is introduced in Section 12. This provides a framework for reviewing the evidence from several studies, including the statistical technique of **meta-analysis**. In Section 13, you will work through an article published in the medical literature. This uses ideas and techniques from all three parts of this book.

10 Randomized controlled trials

A randomized controlled trial, or RCT, is a particular type of cohort study. The main, and crucial, difference between RCTs and the cohort studies that you studied in Parts I and II is that the investigator has much more control over their design. This means that bias and confounding can be kept to a minimum.

The experimental plan of a randomized controlled trial is set out in a document called the **trial protocol**. This describes the rationale for the trial and sets out in detail how the trial is to be organized. For example, the protocol will describe who is eligible to take part and how many participants are required, how they will be allocated to the treatment group and the control group, how they will be followed up, how the analysis will be undertaken, and how safety of the treatment is to be monitored.

Randomized controlled trials are an extremely powerful tool and occupy a central place in modern medical statistics. Whole books have been written about their design, conduct and analysis. The purpose of this section is to provide an introduction to some of the features of randomized controlled trials that distinguish them from other types of cohort studies. Thus randomization, concealment, and analysis are discussed in Subsections 10.1, 10.2 and 10.3, respectively. The role of RCTs in drug development is described briefly in Subsection 10.4.

10.1 Randomization

As for other cohort studies, a randomized controlled trial typically involves an exposure group and a control group. (Some RCTs include several exposure groups and control groups.) In the context of RCTs, the exposure group is usually called the **treatment group**, or the **intervention group**, depending on the context. The purpose is to compare the frequencies of a given disease outcome in the treatment group and the control group — for example, by estimating the odds ratio.

The single most important difference between RCTs and other, non-randomized, cohort studies is that, after a person has been enrolled in the study, their allocation to the treatment group or the control group is decided by chance alone, using a procedure known as **randomization**. In a trial with one treatment group and one control group, each participant is typically allocated to the treatment group with probability $\frac{1}{2}$ and to the control group with probability $\frac{1}{2}$. In its simplest form, the allocation could be decided by the toss of a coin: allocate a newly recruited participant to the treatment group if the coin lands Heads uppermost, and to the control group if it lands Tails uppermost. The patients allocated in this way are said to have been **randomized** to the treatment group and the control group.

In modern RCTs, the randomization is usually carried out using a computer rather than by tossing a coin.

Example 10.1 Early aspirin use in patients with stroke

In the 1980s and 1990s, large RCTs demonstrated that taking aspirin was effective in reducing mortality in patients with heart disease and stroke. One of these trials was undertaken in China in people who had suffered a stroke, the aim being to investigate whether early treatment with aspirin improved survival.

The trial included 21 106 patients in two groups: 10 554 received aspirin, and 10 552 received a placebo (an inactive substance). After being recruited, patients were allocated to treatments at random. To do this, a randomization list was prepared, consisting essentially of a sequential list of patient numbers, and a list of treatment codes in random order. Part of such a randomization list, which allows for a total 21 110 patients, is shown in Table 10.1 (overleaf).

Chen, Z.M., Collins, R., Liu, L.S., Pan, H.C., Peto, R. and Xie, J.X. (1997) CAST: randomised placebo-controlled trial of early aspirin use in 20 000 patients with acute ischaemic stroke. *Lancet*, **349**, 1641–1649.

Envelopes carrying the sequential patient numbers were packed with either aspirin or placebo according to the randomization list, sealed, strung together through punched holes to ensure they would be used in sequence, and dispatched to the 413 hospitals in China that participated in the trial.

A stroke patient recruited to the trial was given whichever treatment was included inside the next available envelope. ♦

Table 10.1 Part of a randomization list

Patient number	Treatment
00001	Aspirin
00002	Placebo
00003	Placebo
00004	Aspirin
00005	Placebo
⋮	⋮
21109	Aspirin
21110	Aspirin

Randomization has two main aims. The first is to remove selection bias and confounding. In the absence of random allocation, the investigators might tend to allocate patients to treatment or control according to the individual patient's characteristics or prognosis — for example, by allocating the sicker patients to the treatment group. Even if this does not occur, the suspicion might remain that it did, and hence the results of the trial might carry less conviction. Randomization removes any overt or unwitting selection bias that might occur in allocating patients to treatment or control.

Randomization also produces an approximate **balance** in the characteristics of the patients allocated to the treatment group and the control group, so that the two groups are comparable. For example, suppose that the participants in a trial include 75% men and 25% women, and that the participants are randomized to two groups in equal proportions. Then about half the males and half the females will be allocated to the treatment group. Equivalently, about 75% of those randomized to each group will be men: the groups are said to be balanced with respect to sex. Note that such balance is only approximate, since there will be differences between the groups attributable to random variation. However, on average, and in large samples, the treatment group and the control group will be comparable.

Balance is important because it removes confounding. To see this, suppose that age is believed to be a potential confounding variable, because it is associated with the outcome. To introduce confounding bias, age must also be associated with the treatment. However, if the treatment groups are balanced with respect to age, the individuals allocated to the treatment group and the control group have a similar age distribution. Hence treatment is not associated with age, and thus age cannot act as a confounder.

You learned about selection bias and confounding in Section 6.

A second, more technical, reason why randomization is a good thing is that it ensures that statistical tests and confidence intervals used to report the trial are soundly based in theory. In most non-randomized studies, it is hoped that the theory applies, but there is no guarantee that it does. Randomization ensures that the theory does apply.

Precisely why this is so is beyond the scope of this course. But here is a taster: the repeated experiments interpretation of confidence intervals can be proved to work for RCTs because re-randomizing the same patients can be thought of as 'repeating the experiment'.

Example 10.2 Balance in the Chinese aspirin trial

In the Chinese trial of aspirin against placebo in stroke patients described in Example 10.1, good balance between the two groups was achieved in the characteristics of the patients at the time they were recruited, as shown in Table 10.2.

Table 10.2 Characteristics of patients at recruitment

Characteristic	Aspirin group ($n = 10\,554$)	Placebo group ($n = 10\,552$)
Male	6633 (62.8%)	6734 (63.8%)
Aged ≥ 70	2938 (27.8%)	2948 (27.9%)
Time from stroke > 24 hours	4906 (46.5%)	4932 (46.7%)
Blood pressure < 160 mm Hg	5460 (51.7%)	5434 (51.5%)

Note that the proportions differ between the groups, due to random fluctuations in the allocation. However, these differences are small and have no bearing on the results. ♦

In theory, randomization should produce balanced groups, at least in large trials. However, it is often sensible to build balance more explicitly into the design rather than relying on chance. This is the case, for example, for trials carried out in several different places. Such trials are called **multi-centre trials**. In multi-centre trials it is desirable to achieve balance within each participating centre. To take an extreme example, suppose that a small trial is to be undertaken in two centres A and B. If, by chance, the randomization allocates most of the patients in centre A to the treatment group and most of the patients in centre B to the control group, the effect of the treatment would be confounded by centre: any differences between treatments could also be attributed to differences between centres.

More generally, if the number of participants in each centre is small, randomization might not achieve good balance within centres. This may be remedied by employing **stratified randomization** within centres. The usual way to do this is to **randomize by blocks**. The procedure is described in Example 10.3.

Example 10.3 Stratified randomization in the Chinese aspirin trial

The Chinese aspirin trial, which was described in Example 10.1, was conducted in 413 centres. To make sure that in each centre the numbers in the aspirin group and the placebo group were roughly similar, the randomization list in Table 10.1 was constructed in such a way that, of the first ten patient numbers, five were allocated to aspirin and five to placebo. In the same way, five of the patient numbers 11 to 20 were allocated to aspirin, and five to placebo. Similarly, for each further block of ten patient numbers — 21 to 30, 31 to 40, and so on — five were allocated to aspirin and five to placebo. This is an example of randomization by blocks with **block size** 10. Within each block, the allocation is random, but the total allocated to each group is known for every block. In this trial, the envelopes (see Example 10.1) were sent in blocks of ten to each study site. For each block of ten, five patients were allocated to aspirin and five to placebo. Thus any difference between the numbers allocated to placebo and aspirin within each centre comes from an incomplete final block, and such a difference cannot be more than five. ♦

Stratified randomization may also be used to ensure that variables known to be associated with the outcome are approximately balanced — for example, within age groups, or within groups of patients with common prognostic factors. One way to achieve this is to prepare a separate randomization list for each grouping, or stratum, and to randomize by blocks within each stratum.

Activity 10.1 Guillain–Barré syndrome

Guillain–Barré syndrome is a major cause of acute paralysis in developed countries. While most patients recover spontaneously, some symptoms, including complete paralysis of the lower limbs, may persist. The risk of this occurring varies with age, being more frequent in patients aged 50 years or more. A randomized controlled trial is planned of a new treatment for Guillain–Barré syndrome, compared to the standard treatment. The outcome variable is the long-term outcome of the disease. It is anticipated that about 200 patients will be enrolled in the trial.

It is decided to stratify the randomization by the patient's age, in two strata — < 50 years and ≥ 50 years. Accordingly, a separate randomization list is prepared for each age group. Each list uses block randomization, in blocks of six, and patients are allocated to treatment using the randomization list corresponding to their age group.

(a) What is the largest possible difference between the numbers of patients allocated to the two treatments within each age group?

(b) Explain briefly why it is advisable to stratify the randomization by age group.

So far it has been assumed that the trial includes two groups of (roughly) the same size. Trials with more than two groups can also be randomized by blocks. For example, if there are to be three groups of (roughly) equal size, the block size should be divisible by 3. In a randomization with block size 12, for example, each block will contain four patient numbers allocated to each of the three treatments, in random order.

Activity 10.2 Randomization by blocks with more than two groups

It is required to design a trial with four groups of roughly equal size within each participating centre. Describe how this might be achieved using randomization by blocks, and select an appropriate block size.

Other forms of randomization may also be used. For example, in **cluster randomized** trials, whole groups of individuals (such as schools or cities) are randomized. This is necessary when an intervention can only be applied to groups of individuals. Examples of such interventions include health promotion campaigns (which may be conducted within a school, for instance) and some collective interventions such as water fluoridation (in which whole cities are randomized to fluoridation or control). Cluster randomized trials require special methods of analysis, and will not be considered further in this book.

10.2 Concealment

In Subsection 10.1, you learned that one purpose of randomization is to eliminate selection bias. Information bias, on the other hand, can be eliminated (or at least reduced) by concealing the allocation to treatment group or control group. Such concealment procedures in randomized controlled trials are known as **blinding**.

Information bias was discussed in Subsection 6.3.

In some circumstances, if trial participants, or those responsible for collecting data on trial participants, know to which group they have been allocated, this knowledge could influence their judgement, consciously or unconsciously, about the success or otherwise of the intervention. This may in turn result in information bias.

Ideally, neither the trial participants nor the persons responsible for collecting data on the trial participants should know which person has been allocated to which treatment. If this is the case, the trial is called **double-blind**: the allocation is concealed from both trial participants and trial staff. An example of a double-blind trial is given in Example 10.4.

Example 10.4 Trial of whooping cough vaccine in the UK

In the late 1940s a randomized controlled trial was undertaken in the UK to evaluate the efficacy of a vaccine against whooping cough. After enrolment, participating children were randomized, and about half received the whooping cough vaccine. They were then followed up for 27 months on average. Each month, the child was visited by a nurse who took details of any symptoms consistent with whooping cough. (Symptoms include coughing with characteristic 'whooping', vomiting, and high temperature.) On the basis of these observations, children were classified as either having had whooping cough or not.

Medical Research Council (1951) The prevention of whooping-cough by vaccination. *British Medical Journal*, 30 June, 1463–1471.

Inevitably a degree of subjective judgement enters into the classification of children as cases of whooping cough. Such judgements may be affected by knowledge of whether or not the child had received vaccine. For example, a borderline case might be more likely to be classified as whooping cough if the nurse knew that the child had not received the vaccine. Similarly, the symptoms might have been reported differently by the child and the child's parents if they knew to which group the child had been allocated.

In such circumstances, information bias would undoubtedly be a problem. To eliminate (or at least substantially reduce) this bias, the allocation of children was concealed from participants and investigators. This was a double-blind trial.

Double blinding was achieved by allocating a placebo vaccine to the control group. In every way the placebo vaccine looked identical to the whooping cough vaccine, but of course it did not protect against whooping cough. The randomization code was kept secret until the end of the trial. ♦

To maintain concealment, individuals randomized to the control group must receive a placebo that looks like the active treatment. In some instances concealment is not possible, in which case the trial is called an **open trial**. Such a trial is described in Example 10.5.

Example 10.5 Open trial of counselling for mothers considering breast feeding

Breast feeding of newborn babies improves the health of both baby and mother. However, in the UK, only 69% of babies born in 2000 were initially breast fed. In order to assess the impact of voluntary counselling to promote breast feeding, a randomized trial was undertaken.

The participants in this trial were pregnant women considering whether or not to breast feed their child. Women were randomly allocated to two groups, intervention and control. Women in the intervention group received counselling about breast feeding, through home visits, postnatal support and information leaflets. Women in the control group received standard care. The outcome was whether or not the mother was breast feeding her baby six weeks after birth.

In this trial it is clearly not possible to arrange any concealment: the very nature of the intervention precludes it. The trial is therefore an open trial. ♦

Graffy, J., Taylor, J., Williams, A. and Eldridge, S. (2004) Randomised controlled trial of support from volunteer counsellors for mothers considering breast feeding. *British Medical Journal*, **328**, 26–29.

Double-blind trials and open trials represent the two extremes in concealment. Double blinding is strongly recommended whenever possible. In some circumstances, full double blinding is not feasible, but some degree of concealment is nevertheless possible. An example is given in Activity 10.3.

Activity 10.3 Yellow oleander poisoning in Sri Lanka

In Sri Lanka, some 2000 people die each year from poisoning due to ingestion of seeds of the yellow oleander. The standard treatment includes drinking activated charcoal dissolved in water. A trial was organized to investigate whether repeating the charcoal treatment would reduce death rates further.

Patients were randomized to two groups. In the control group, patients drank the standard dose of dissolved charcoal, and thereafter drank pure water. In the treatment group, patients drank the standard dose of dissolved charcoal, and thereafter drank further doses of dissolved charcoal.

De Silva, H.A., Fonseka, M.M.D., Pathmeswaran, A. *et al.* (2003) Multiple-dose activated charcoal for treatment of yellow oleander poisoning: a single-blind, randomised, placebo-controlled trial. *Lancet*, **361**, 1935–1938.

(a) No placebo was available of the same appearance as dissolved charcoal. In these circumstances, is it possible to conceal from the patients which treatment they received? Why was water given to patients in the control group?

(b) In this trial, the doctors evaluating the patients did not know to which group the patients had been allocated. How might this have been arranged?

The trial described in Activity 10.3 is a **single-blind** trial: in this trial, the allocation was concealed from the doctors evaluating the patients, but not from the participants. However, in some single-blind trials, the allocation is concealed from the patients but not from the investigators.

10.3 Analysis of randomized controlled trials

Most of the work involved in conducting a randomized controlled trial lies in its design and implementation. The analysis, on the other hand, is usually a relatively simple matter. For example, randomization should ensure approximate balance between groups, and hence eliminate confounding. Balance should of course be checked, to ensure that the randomization has worked. It is possible, though unlikely, that there will be serious imbalances between treatment groups after randomization. But, if this is the case, then more complicated methods of adjustment such as those described in Section 7 may be needed.

One important issue in the analysis of RCTs is deciding which trial participants to include in the analysis. The issue is illustrated by Example 10.6.

Example 10.6 Pelvic exercise and foetal position at birth

Most births occur with the baby emerging head first. However, in a small number of cases, the baby is born bottom or feet first. This is called a breech birth, and is associated with increased pain in labour, and increased chance of complications afterwards. Since the 1950s, it has been suggested that pelvic exercises by the mother prior to birth increases the chances of the baby being born head first. A randomized controlled trial was undertaken to evaluate this recommendation. Pregnant women were allocated randomly to daily pelvic exercises (the intervention group) or to a routine of daily walking (the control group): 1292 women were recruited to the intervention group, and 1255 to the control group. Of these women, 105 in the intervention group and 98 in the control group had a breech birth. The data are in Table 10.3.

Kariminia, A., Chamberlain, M.E., Keogh, J. and Shea, A. (2004) Randomised controlled trial of effect of hands and knees posturing on incidence of occiput posterior position at birth. *British Medical Journal*, **328**, 490–493.

Table 10.3   Outcomes in women randomized to intervention or control

Trial group	Breech births	Other births	Total randomized
Intervention	105	1187	1292
Control	98	1157	1255

The proportion of breech births in the intervention group is $105/1292 \simeq 0.081$; in the control group, the proportion is $98/1255 \simeq 0.078$. The estimated odds ratio for the association between pelvic exercise and breech birth is 1.04, with 95% confidence interval $(0.78, 1.39)$.

However, during the course of the trial, 246 of the women randomized to the intervention group withdrew or had early labour, and hence did not complete the course of pelvic exercises. In the control group, 46 women withdrew or had early labour. It might be argued that only participants who completed the trial should be evaluated. Of those who completed the trial, 82 in the intervention group and 96 in the control group had breech births. The data are in Table 10.4.

Table 10.4   Outcomes for women who completed the trial

Trial group	Breech births	Other births	Total remaining
Intervention	82	964	1046
Control	96	1113	1209

With these data, the proportion of breech births in the intervention group is $82/1046 \simeq 0.078$, and the proportion in the control group is $96/1209 \simeq 0.079$. The estimated odds ratio is 0.99, with 95% confidence interval $(0.73, 1.34)$.

Tables 10.3 and 10.4 thus present two different approaches to analysing the trial data. While neither suggests that pelvic exercise has any appreciable effect on foetal birth position, they produce slightly different results. Such differences may matter more in other contexts, and raise the question 'which approach should be used in the final analysis of the trial?' ♦

Two ways of analysing the data are presented in Example 10.6. The first approach is to use all individuals randomized, irrespective of whether or not they completed the treatment to which they were allocated. This method of analysis is called analysis by **intention to treat**: participants are counted in the treatment group to which they were allocated by randomization, even if they later switch treatments. The second approach is to include only those participants who complete the treatment to which they were randomized. This method of analysis is called **per protocol**.

The analysis by intention to treat is recommended. The reason is that switching treatments and other departures from the protocol (collectively referred to as **protocol violations**) may be associated with the treatment, thus introducing a selection bias among those individuals who complete the trial strictly according to the protocol. Example 10.7 illustrates how such a bias might arise.

Example 10.7 Selection bias after randomization

Suppose that a new treatment is to be evaluated against the standard therapy in a randomized controlled trial and that 100 individuals are randomized to each group. The proportions of patients who have improved after six months are to be compared.

In the control group receiving standard therapy, all 100 participants finish the trial, and 10 are found to have improved after six months. The new treatment, on the other hand, is found to induce some side effects, and many patients have switched back to the standard therapy owing to these side effects. However, those who switched back to the standard therapy tend to be primarily those who felt they were not experiencing any benefit from the treatment. Suppose that, out of the 100 originally randomized to the new treatment, 60 remained on the new treatment for the full six months. Of these, 20 improved and 40 did not. Of the 40 who switched back to the standard therapy, 5 improved and 35 did not.

In an intention-to-treat analysis, the success rate for the new treatment is $25/100 = 0.25$, and for the standard therapy it is $10/100 = 0.10$. The relative risk of success is thus $0.25/0.10 = 2.5$.

In contrast, in a per-protocol analysis, only those who stay on the treatment to which they were allocated are included. In this case, the success rate in the new treatment group is $20/60 \simeq 0.33$, and as before, for the standard therapy the success rate is 0.10. The relative risk is now $0.33/0.10 = 3.3$, which is greater than the value (2.5) obtained by intention to treat.

However, the denominator for the new treatment group in the per-protocol analysis excludes 40 who dropped out partly because they had experienced no improvement up to that point. This introduces a selection bias that makes the treatment look better than it really is. The intention-to-treat analysis, on the other hand, uses denominators that are created by a purely random mechanism, and hence are unaffected by selection bias. ♦

The purpose of an intention-to-treat analysis is to keep to (or as close as possible to) the treatment allocation decided by the randomization, so as to avoid subsequent selection bias. Admittedly, the intention-to-treat analysis is not always wholly satisfactory. In the pelvic exercise trial of Example 10.6, it means that the group of women randomized to treatment includes some who did not in fact complete the course of exercises.

In some cases it is not possible to analyse the groups exactly as randomized — for example, if some participants have withdrawn completely and their outcomes are not known. In Example 10.6, this was not a problem since the foetal position of the children born to all women enrolled was known, whether or not they withdrew from the study. But in other cases this might not be so, and losses may be inevitable. As a general principle, an intention-to-treat analysis should attempt to keep as close as possible to the groups that were randomized at the outset.

For this reason, much effort is expended in RCTs to reduce the numbers of participants who drop out or switch groups, and other protocol violations. However, in most trials some losses and errors inevitably occur. It is good practice, in reporting randomized controlled trials, to draw up a flow chart showing the numbers included and excluded at each stage of the trial and to document the reasons why participants were excluded or lost to follow-up.

Example 10.8 *Flow chart for the pelvic exercise and foetal position trial*

Figure 10.1 shows the flow chart for the trial of pelvic exercise and foetal position described in Example 10.6.

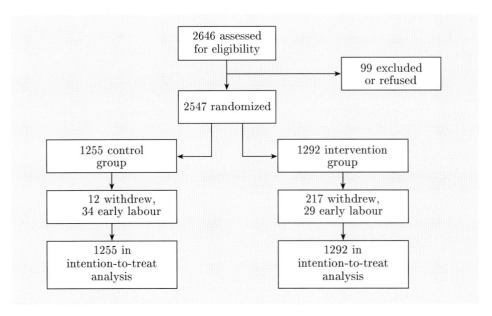

Figure 10.1 Flow chart for pelvic exercise trial

The top box states that 2646 women were assessed for eligibility. Of these women, 43 declined to take part and 56 did not meet the criteria for inclusion (for example, women carrying twins were excluded). This left 2547 women who were randomized to the intervention group (1292 women) or the control group (1255 women). Of these, 229 withdrew (that is, decided to stop participating) and 63 went into early labour, and hence did not complete the course of exercises to which they were allocated. All 2547 women randomized were included in the analysis, as data were collected on the foetal positions at birth of all babies. ♦

Activity 10.4 Analysis of the yellow oleander poisoning trial

In Activity 10.3, a trial of supplementary charcoal therapy to treat yellow oleander poisoning was described. Figure 10.2 shows the flow chart for the trial.

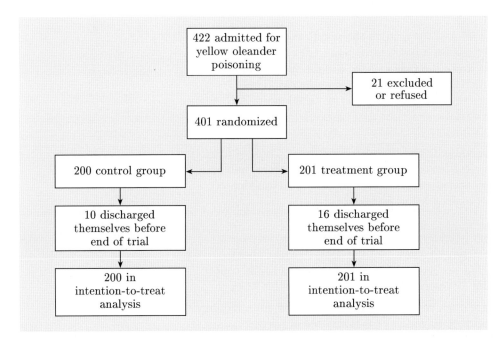

Figure 10.2 Flow chart for the yellow oleander poisoning trial

Of the 200 randomized to the control group, ten discharged themselves before the end of the trial and of the remaining 190, sixteen died. In the treatment group, sixteen discharged themselves before the end of the trial and of the remaining 185, five died. Those who discharged themselves are treated as survivors. Carry out an intention-to-treat analysis of this trial, as follows.

(a) Draw up a table showing the data to be included in the intention-to-treat analysis.

(b) Using an intention-to-treat analysis, estimate the odds ratio for the association between supplementary charcoal therapy and death from poisoning, and calculate a 95% confidence interval.

(c) What do you conclude about the effect of supplementary charcoal therapy in the treatment of yellow oleander poisoning?

10.4 Evaluation of pharmaceutical drugs

In Subsections 10.1 to 10.3, some features common to all randomized controlled trials have been described. RCTs can be undertaken to evaluate a wide range of interventions in the medical field and beyond. In this subsection, you will learn a little more about the application of RCTs in one particular field — the evaluation of pharmaceutical drugs.

The first randomized controlled trial is believed to be the 1948 trial of the antibiotic streptomycin in the treatment of tuberculosis in the UK. Certainly, RCTs only became the accepted method for evaluating drugs after 1950, largely under the influence of Austin Bradford Hill. Today, randomized controlled trials form a key component of the evaluation procedure, and evidence of efficacy and safety from such trials is necessary before a drug can be licensed for commercial use.

The procedure for evaluating a new drug in humans follows four phases.

- **Phase I studies**
  These are the first studies of the drug in humans. They are small, involve adult volunteers (often some of the scientists who developed the drug), and are focused mainly on testing the safety and biological action of the drug. Some Phase I studies may not include any controls.

- **Phase II studies**
  These are usually the first studies of the drug in the population for whom it was intended. They typically involve a few tens or hundreds of patients. Their main aim is usually to study the safety of the drug in more detail, and to investigate the biological action of the drug at different doses. These studies are usually controlled, and are often designed as randomized controlled trials.

- **Phase III studies**
  These are larger studies to investigate the efficacy of the drug. These studies are always randomized controlled trials, and typically involve several hundred or thousand participants (sometimes many more). Successful completion of Phase III trials may lead to the drug being licensed.

- **Phase IV studies**
  After the drug is licensed and in widespread use, further studies may be carried out to monitor the safety and efficacy of the drug. Such studies can be much larger than RCTs, but are not randomized. They are usually cohort studies or case-control studies as described in Parts I and II of this book.

Studies undertaken on patients are often called **clinical trials**. All studies require ethical approval by suitable authorities to ensure that the rights of participants are safeguarded and that the trial is conducted according to ethical principles.

Activity 10.5 A new vaccine against typhoid fever

A study was undertaken to evaluate a new candidate vaccine against *Salmonella typhi*, the bacterium that causes typhoid. Three healthy adult male volunteers were given the vaccine and carefully monitored for two weeks for adverse reactions and for the biological action of the vaccine.

From this brief description, state whether this study should be classified as a Phase I, II, III or IV study, and identify three reasons to support your choice.

Bellanti, J.A., Zeligs, B.J., Vetro, S. *et al.* (1993) Studies of safety, infectivity and immunogenicity of a new temperature-sensitive (ts) 51–1 strain of *Salmonella typhi* as a new live oral typhoid fever vaccine candidate. *Vaccine*, **11**, 587–590.

Randomized controlled trials of new drugs are experiments on humans, and hence are governed by strict rules, and require careful monitoring. If a new treatment that is being evaluated in a randomized controlled trial turns out to have serious side effects, the trial may have to be stopped early to avoid harming any more patients. Similarly, if a drug evaluated against a placebo turns out to be highly effective, it may be unethical to continue administering the placebo. In this case also, the trial should be stopped and all patients should be given the new drug.

However, a decision to stop early should not be taken lightly. For example, the drug might be effective for a short period, then lose its efficacy. If the trial is stopped early, this effect may not be noticed.

In many trials the decision to continue the trial or to stop early is taken by an independent Data Monitoring Committee. This committee undertakes confidential interim analyses of the data — that is, analyses before the trial is finished — and recommends whether or not the trial should continue. The committee members have full access to the data, unblinded if necessary, and have the authority to stop the trial if they decide that ethical considerations require it. This committee usually includes a statistician, who provides advice on the strength of the evidence underpinning the committee's recommendation.

Activity 10.6 Trial of hormone replacement therapy after breast cancer

In 1997, an open randomized controlled trial was started to investigate the treatment of menopausal symptoms in women who had previously had breast cancer. Women were randomized to two groups: hormone replacement therapy (HRT), and no HRT.

A Data Monitoring Committee was set up to review regularly the results of the trial. All analyses were to be done by intention to treat. By the third meeting of the Committee, 434 women had been randomized. Follow-up data were available on 345 women. Of these, 34 had experienced new breast cancers. The data are in Table 10.5.

Holmberg, L. and Anderson, H. (2004) HABITS (hormonal replacement therapy after breast cancer – is it safe?), a randomised comparison: trial stopped. *Lancet*, **363**, 453–455.

Table 10.5 Interim data from HRT trial

Trial group	Total randomized	Women with data available	New breast cancer
HRT	219	174	26
No HRT	215	171	8

(a) Of the women randomized to HRT, 11 did not actually receive HRT. Of the women randomized to no HRT, 39 did in fact receive HRT. What data would you use in an intention-to-treat analysis? Draw up a suitable table.

(b) Carry out the intention-to-treat analysis to estimate the odds ratio for the association between HRT and new breast cancer, and calculate a 95% confidence interval.

(c) If you were a member of the Data Monitoring Committee, what would you recommend, and why?

Summary of Section 10

Randomized controlled trials are experimental cohort studies to evaluate a treatment or intervention. Participants are randomly allocated to exposure groups, a procedure known as randomization. This removes selection bias, ensures that the treatment groups are approximately balanced, and provides a sound theoretical basis for the statistical analysis. The random allocation may be stratified, using randomization by blocks. To remove information bias, the allocation to exposure groups may be concealed, a procedure known as blinding. This may take the form of double blinding or single blinding. To make sure that the analysis is not affected by selection bias, the analysis should be undertaken based on the intention-to-treat principle — as far as possible, individuals should be analysed according to the groups to which they were randomized. The testing of a new drug in human subjects involves four phases. Phase III studies (and some Phase II studies) are randomized controlled trials to evaluate efficacy and safety.

Exercises on Section 10

Exercise 10.1 Trial of pneumococcal vaccine against otitis media

Otitis media is an inflammation of the middle ear that can, if untreated, result in glue ear and other permanent damage. Many cases of otitis media occur in children, and are often caused by infection by the bacterium *Streptococcus pneumoniae*. A vaccine has been developed against this bacterium. A trial was organized in the Netherlands to find out whether it could be used to reduce the incidence of otitis media in children aged between 1 and 7 years.

The trial was designed with two groups of equal size: the vaccine group, and a control group (who received another vaccine, which was known not to have any effect on otitis media). The organizers considered that it was important to maintain balance with respect to age (grouped as 12–24 months and 25+ months) and previous incidence of otitis media (grouped as low and high). Briefly describe how you might organize the randomization to achieve such balance.

Veenhoven, R., Bogaert, D., Uiterwaal, C. *et al.* (2003) Effect of conjugate pneumococcal vaccine followed by polysaccharide pneumococcal vaccine on recurrent acute otitis media: a randomised study. *Lancet*, **361**, 2189–2195.

Exercise 10.2 Analysis of the pneumococcal vaccine trial

This exercise is based on the pneumococcal vaccine trial described in Exercise 10.1.

In the trial, 388 children were recruited, of whom 383 were randomized — 190 to the treatment group and 193 to the control group. The trial flow chart is shown in Figure 10.3.

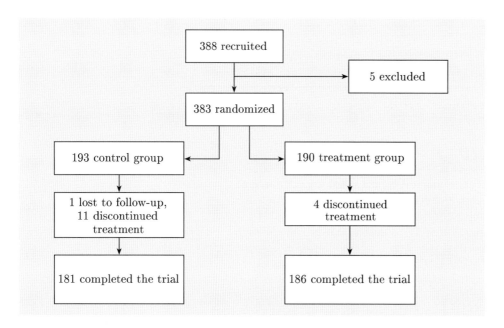

Figure 10.3 Flow chart for the pneumococcal vaccine trial

The assessment period lasted for eighteen months, starting one month after receiving the third dose of vaccine. During the assessment period, 107 children in the treatment group and 101 children in the control group experienced one or more episodes of acute otitis media.

(a) Draw up the data table for an intention-to-treat analysis of the association between vaccination and occurrence of at least one otitis media episode during the trial. Outcomes were available on all children randomized, with the exception of one child lost to follow-up.

(b) Calculate the odds ratio and a 95% confidence interval.

(c) What do you conclude about the efficacy of pneumococcal vaccination in reducing the proportion of children who experience otitis media?

11 Choosing the sample size

In this section, the issue of choosing the sample size for a randomized controlled trial (that is, the number of participants to be randomized) is considered from the statistical point of view. The methods described are applicable to cohort studies more generally. However, design issues, including the choice of sample size, are particularly important for randomized controlled trials. One of the first tasks when designing an RCT is to specify the trial hypotheses. Hypotheses are discussed in Subsection 11.1, and the idea of the power of a fixed-level test is introduced. In Subsection 11.2, a formula is given for the sample size to be used in a two-group trial. The use of this formula is discussed in Subsection 11.3, and the power available in a trial with groups of a given size is calculated.

11.1 Trial hypotheses, significance level and power

The calculation of the sample size required in a randomized controlled trial is derived within the framework of hypothesis testing called **fixed-level testing**. The trial hypotheses are the null hypothesis, denoted H_0, and the alternative hypothesis, denoted H_1. Suppose that a two-group trial is being designed to investigate the effect of a treatment or intervention on a disease D. The proportions of patients allocated to the treatment group (T) and the control group (C) with the disease D are to be compared. Let

$$p_T = P(D|\text{treatment}), \quad p_C = P(D|\text{control}).$$

The motivation for the trial is usually that there is some evidence, from laboratory experiments or other trials, that the treatment may have a beneficial effect. In statistical terms, the trial is designed to reject the null hypothesis H_0 that the treatment has no effect in favour of the alternative hypothesis H_1 that the treatment has an effect. These hypotheses may be written as

$$H_0 : p_T = p_C, \quad H_1 : p_T \neq p_C.$$

This formulation defines a **two-sided hypothesis test**. The test is two-sided because the alternative hypothesis does not specify the direction of the effect — it includes $p_T > p_C$ and $p_T < p_C$. In some circumstances, one-sided alternative hypotheses are used — for example, $H_1 : p_T > p_C$ or $H_1 : p_T < p_C$. In this course, only two-sided alternative hypotheses are considered.

Example 11.1 Statins and heart disease

Statins are a class of drugs that reduce the level of harmful cholesterol in the blood, and hence may reduce the risk of heart disease in some individuals. Many clinical trials of statins have been undertaken in recent years. In one such trial, the effect of statins was investigated in UK adults with coronary or other types of heart disease or diabetes and aged between 40 and 80 years.

Collins, R., Armitage, J., Parish, S., Sleight, P. and Peto, R. (2002) MRC/BHF Heart Protection Study of cholesterol lowering with simvastatin in 20 536 high-risk individuals: a randomised placebo-controlled trial. *Lancet*, **360**, 7–22.

The main outcome of interest was death, from any cause, over a period of five years. The trial included two groups: a treatment group, comprising individuals randomized to receive simvastatin (a particular type of statin), and a control group, comprising individuals randomized to receive placebo. In this case, p_C is the probability of dying within five years after randomization for an individual allocated to placebo, and p_T is the probability of dying within five years after randomization for an individual allocated to simvastatin.

The null hypothesis is $H_0 : p_T = p_C$, meaning 'the underlying five-year risk of death from any cause is the same in individuals allocated to receive simvastatin as it is in individuals allocated to receive placebo'. The alternative hypothesis is $H_1 : p_T \neq p_C$, meaning 'the underlying five-year risk of death from any cause is not the same in individuals allocated to receive simvastatin as it is in individuals allocated to receive placebo'. ◆

When it comes to formulating hypotheses, there is sometimes some confusion as to whether 'the hypothesis' refers to H_0 or H_1. (We have avoided this pitfall by referring to 'the trial hypotheses' in plural.) The confusion arises because of the rather special way in which statisticians think of hypothesis testing. This is illustrated in Example 11.2.

Example 11.2 Formulating hypotheses

The clinical trial of statins described in Example 11.1 was undertaken in a group of patients at high risk of heart disease. Previous experience with statins suggested that they may reduce the risk of heart disease in such patients, but conclusive evidence in this particular group of patients was lacking.

Thus, from a scientific viewpoint, there was good reason to suppose, prior to the trial, that p_T, the proportion of individuals given statins who die within five years, should be less than p_C, the proportion of individuals not given statins who die within five years. A medical scientist might thus describe the purpose of the trial as being

> 'to test the hypothesis that simvastatin reduces the five-year death risk'.

Written more formally, this could be expressed as a one-sided alternative hypothesis $p_T < p_C$. From a statistical perspective, however, hypotheses cannot be proved, only refuted. Thus a statistician would describe the purpose of the trial as

> 'to test the hypothesis that simvastatin has no effect on the five-year death risk',

that is, to weigh up the evidence against the null hypothesis $p_T = p_C$. Rejecting the null hypothesis provides evidence in favour of the alternative hypothesis $p_T \neq p_C$. The strength of the association, and its direction, can then be quantified using odds ratios and confidence intervals.

In the above two descriptions of the purpose of the trial, the scientist and the statistician use the word 'hypothesis' in slightly different ways. For the scientist, the term hypothesis relates to the evidence or beliefs that the drug will work that are available prior to the trial. This prior hypothesis, denoted H_P, can be expressed as follows:

$H_P : p_T < p_C$.

Such a prior hypothesis, preferably backed by evidence, is essential in providing the rationale for the trial. Without it, ethical approval for the trial is unlikely to be granted. (Conversely, if the prior evidence is too strong, then the trial is unnecessary and ethical approval may also be withheld.)

The statistician uses the term 'hypothesis' in a more formal sense than the scientist, to define a framework in which to obtain and evaluate evidence relating to H_P. The *statistical hypotheses* are

$H_0 : p_T = p_C, \quad H_1 : p_T \neq p_C$.

As you will see in Subsection 11.2, calculating the sample size for a trial requires both points of view. ♦

Activity 11.1 will give you some practice at specifying the trial hypotheses, on the basis of an informal description of the trial rationale.

Activity 11.1 Infliximab and rheumatoid arthritis

In recent years, several effective new treatments against rheumatoid arthritis have been developed. While these treatments do not cure the disease, they may reduce the severity of symptoms, as measured on a scale developed by the American College of Rheumatology (the ACR scale).

Section 11 Choosing the sample size

One new drug is called Infliximab. A trial is to be undertaken of standard therapy plus Infliximab against standard therapy plus a placebo. The outcome of interest is the ACR50, the proportion of patients whose condition improves by 50% or more on the ACR scale over a period of a year. The prior evidence is that Infliximab is likely to increase the ACR50.

(a) What are the treatment group and the control group in this trial?

(b) Formulate the trial hypotheses clearly.

Owing to random fluctuations, it is not possible to guarantee that a trial will yield the 'correct' result. For example, it may be the case that the treatment has no effect on average, but that, by chance, in the sample selected in the trial, the individuals allocated to the treatment group do much better or much worse than those in the placebo group. In this case it may be concluded, wrongly, that the null hypothesis is false. An error of this type is called a **Type I error**. The probability of a Type I error is thus defined by

Type I error probability $= P(\text{rejecting } H_0, \text{ given that } H_0 \text{ is true})$.

Another type of error can occur if the null hypothesis is false but, again owing to random fluctuations, similar results are obtained in the treatment group and the placebo group. In this case it is concluded, wrongly, that the null hypothesis should not be rejected. This is called a **Type II error**. The probability of a Type II error occurring is defined by

Type II error probability $= P(\text{not rejecting } H_0, \text{ given that } H_0 \text{ is false})$.

The Type I and Type II error probabilities are related to the properties of the statistical test used to determine whether or not to reject H_0. The Type I error probability is the **significance level** of the test, and is denoted α. Making a Type I error can have very undesirable consequences. In the context of a randomized controlled trial, for example, if an ineffective drug happens by chance to appear effective in the trial, a Type I error will have occurred, which may lead to an ineffective treatment being licensed and used. For this reason, a low significance level is required, usually no more than 5%, and sometimes less.

The Type II error probability is related to the **power** of the test, which is denoted γ. The power is the probability of rejecting H_0, given that it is false, and hence is equal to one minus the Type II error probability:

$$\text{power} = \gamma = 1 - \text{Type II error probability}$$
$$= P(\text{rejecting } H_0, \text{ given that } H_0 \text{ is false}).$$

In the context of a randomized controlled trial of a genuinely effective treatment, the power of the trial is thus the probability that the trial will lead to rejection of the null hypothesis that the drug is ineffective. If the power is low, then there is a high probability that the trial will fail to reject the null hypothesis, even if the drug is effective. One possible consequence is that an effective treatment will not be used, and may be abandoned completely. To avoid such an outcome, a high power is required, typically in excess of 80%.

Activity 11.2 Error probabilities, significance level and power

(a) Calculate the significance level and power of a hypothesis test from the following probabilities:

$P(\text{not rejecting } H_0, \text{ given that } H_0 \text{ is true}) = 0.97,$

$P(\text{not rejecting } H_0, \text{ given that } H_0 \text{ is false}) = 0.10.$

(b) A randomized controlled trial of a new vaccine to prevent infection by HIV, the virus that causes AIDS, is being planned. The outcome of the trial will determine whether the vaccine is recommended for widespread use. Describe briefly what the practical implications of making a Type I and Type II error might be in this context.

The significance level is usually chosen by the investigator, and will determine how results are reported. For example, if a 5% significance level is chosen, then 95% confidence intervals will be quoted. With the significance level fixed in this way, the power depends on the sample size — the larger the sample size, the higher the power. However, if the sample size used in the trial is too small, then the power may be too low to yield useful results.

Example 11.3 Trial of a vaccine against malaria

Malaria is an infectious disease transmitted by some mosquitoes, and is a major cause of death in many tropical countries. Much effort has been devoted to developing a vaccine against malaria. One such candidate vaccine, the SPf66 malaria synthetic vaccine, was evaluated in a randomized controlled trial. In the trial, 230 individuals were allocated to the vaccine group and 238 to the placebo group. During the trial period there were 4 cases of malaria diagnosed within vaccine recipients and 12 within placebo recipients. The significance level for the trial was 5%.

Sempértegui, F., Estrella, B., Moscoso, J. *et al.* (1994) Safety, immunogenicity and protective effect of the SPf66 malaria synthetic vaccine against *Plasmodium falciparum* infection in a randomized double-blind placebo-controlled field trial in an endemic area of Ecuador. *Vaccine*, **12**, 337–342.

The odds ratio for the association between vaccination and malaria is 0.33, with 95% confidence interval (0.11, 1.05). The p value of the test for no association is 0.073 using Fisher's exact test and 0.049 using the chi-squared test.

The evidence of an association is thus rather weak. The confidence interval is very wide because the samples were small, and as a result the true odds ratio cannot be estimated precisely: it could plausibly be as low as 0.11 (indicating a strong protective effect of vaccination) or as high as 1.05 (indicating that vaccination is very slightly *positively* associated with malaria). Thus even if the true odds ratio were much less than 1, we could not conclude that the vaccine works, and hence would make a Type II error. In conclusion, the power of this trial is too low. ♦

In Example 11.3, the power of the study was too low for it to yield conclusive results with a significance level of 5%. The power was too low because the sample size was too small. This example illustrates why choosing a suitable sample size is such an important aspect of the design of a randomized controlled trial. In Subsection 11.2, an expression for the sample size required is given.

11.2 An expression for the sample size

Let p_T and p_C be the underlying proportions with the disease D in individuals receiving the treatment and in individuals receiving the placebo, respectively. Suppose that information on p_T and p_C is to be obtained from a randomized controlled trial involving a total of $2n$ individuals, of whom n will be allocated to the treatment group and n to the placebo group. The aim is to determine n so that the significance level and power take specified values α and γ. In this subsection an expression for the sample size n is given.

The details of the derivation are omitted, but are similar in many respects to those given in *M248 Unit C1 Testing Hypotheses*, Section 5.

It is worth noting from the start that determining the sample size is necessarily a rough and ready procedure: it rests on assumptions and approximations which may or may not be valid. This is inevitable: if everything were known in advance, there would be no need to conduct a randomized controlled trial!

To determine the sample size required, initial estimates π_T and π_C of the true values of p_T and p_C are needed. These initial estimates are called the **design values**. These are usually chosen on the basis of prior evidence or beliefs about their likely values, and on what size of effect $|p_T - p_C|$ is relevant from a practical viewpoint. Some indication of how to choose the design values will be provided in Subsection 11.3.

Section 11 Choosing the sample size

Having chosen the design values, the next step is to define a test statistic S for which the null distribution is approximately $N(0,1)$. For a fixed significance level α, if $q_{1-\alpha/2}$ is the $(1-\alpha/2)$-quantile of $N(0,1)$, then the null hypothesis is rejected if $S < -q_{1-\alpha/2}$ or $S > q_{1-\alpha/2}$. The null distribution of S and the rejection region for the test are shown in Figure 11.1(a).

The definition of S will not be given here.

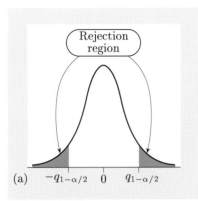
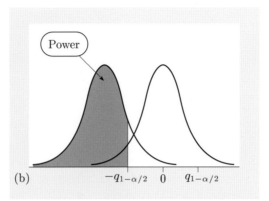

Figure 11.1 (a) The null distribution of S and the rejection region (b) The distribution of S under the alternative hypothesis

Now consider the distribution of S under the alternative hypothesis $H_1 : p_T = \pi_T$, $p_C = \pi_C$. Under the alternative hypothesis, the distribution of S is shifted to one side. This is illustrated in Figure 11.1(b) for the case when the distribution is shifted to the left.

The shaded area in Figure 11.1(b) corresponds to the power of the test. The larger the sample size, the further the distribution of S under the alternative hypothesis shifts either to the left or the right, and hence the greater is the power of the test. The sample size n is chosen so that the power is equal to the specified value γ. An expression for n, the sample size for each of the two trial groups, is given in the following box. The total sample size for the two groups combined is $2n$.

The sample size required for a trial

To calculate the sample size for a trial with two groups, the design values π_T and π_C, the significance level α and the power γ must first be specified. Then the sample size n for each trial group is given approximately by

$$n = \frac{2(q_{1-\alpha/2} + q_\gamma)^2 \pi_0 (1-\pi_0)}{(\pi_T - \pi_C)^2}, \tag{11.1}$$

where $q_{1-\alpha/2}$ and q_γ denote, respectively, the $(1-\alpha/2)$-quantile and the γ-quantile of $N(0,1)$, and $\pi_0 = (\pi_T + \pi_C)/2$.

The required sample size for the trial as a whole is $2n$.

Notice that the sample size required increases as the difference $\pi_T - \pi_C$ decreases in magnitude. Thus the smaller the difference that is to be detected, the larger is the sample size required.

Example 11.4 Calculating the sample size

To calculate the sample size for a trial, all of the following quantities are required: the significance level α, the power γ, and the design values π_T and π_C. Suppose that it is required to design a trial with two groups, with 5% significance level, 90% power, and design values $\pi_T = 0.15$, $\pi_C = 0.2$.

The quantiles $q_{1-\alpha/2}$ and q_γ may be obtained from tables of the standard normal distribution, or using a computer. Some of the values commonly used for sample size calculation are given in Table 11.1.

For $\alpha = 0.05$ and $\gamma = 0.90$, $q_{1-\alpha/2} = q_{0.975} = 1.96$ and $q_\gamma = 1.282$. Also, for $\pi_T = 0.15$ and $\pi_C = 0.2$,

$$\pi_0 = \tfrac{1}{2}(\pi_T + \pi_C) = \tfrac{1}{2}(0.15 + 0.2) = 0.175.$$

Thus, using Formula (11.1), the sample size required in each of the two trial groups is given by

$$\begin{aligned} n &= \frac{2(q_{1-\alpha/2} + q_\gamma)^2 \, \pi_0(1-\pi_0)}{(\pi_T - \pi_C)^2} \\ &= \frac{2 \times (1.96 + 1.282)^2 \times 0.175 \times (1 - 0.175)}{(0.15 - 0.2)^2} \\ &= 1213.97\ldots \\ &\simeq 1214. \end{aligned}$$

Table 11.1 Selected quantiles of the standard normal distribution

α	q_α
0.80	0.8416
0.90	1.282
0.95	1.645
0.975	1.960
0.99	2.326
0.995	2.576

Note that the sample size should be rounded *up* to the next integer, to ensure that the power of the study is not lower than that required. ♦

The expression (11.1) for the sample size per group is derived assuming that no individuals withdraw from the trial. Such losses may reduce the numbers available for analysis, and hence the power of the trial. It is usual to make some estimate of the likely losses to follow-up and adjust the sample size accordingly. If a proportion p are expected to be lost to follow-up, so that the disease outcomes for these individuals are unknown, then the sample size required per group is $n/(1-p)$. Activity 11.3 will give you some practice at calculating sample sizes.

Activity 11.3 More sample sizes

(a) Calculate the total sample size required for a two-group trial with significance level 0.01, power 0.8, and design values $\pi_T = 0.2$, $\pi_C = 0.4$.

(b) Calculate the sample size required per group for a two-group trial with significance level 0.05, power 0.8, and design values $\pi_T = 0.6$, $\pi_C = 0.4$.

(c) Suppose that 10% losses to follow-up are expected in each of the trials described in parts (a) and (b). Calculate the sample sizes required in each trial, allowing for such losses.

11.3 Choosing the sample size

In Subsection 11.2, an expression was given for the sample size that must be used in a two-group trial, given the significance level, power and design values. Given the values of these parameters, calculating the sample size is a purely arithmetic operation. However, choosing a sample size also involves specifying appropriate values of α, γ, π_C and π_T.

The significance level α is usually chosen to be low; typical values are 0.05 or, less commonly, 0.01. The power γ should be chosen to be high: usually a value no less than 0.8 is used. Remember that even with a power of 0.8, there is still a 20% (or 1 in 5) chance of the trial failing to reveal a true effect.

The trial size depends critically on the design values π_T and π_C. The value π_C is the proportion of individuals with the disease event in the placebo or control group. If the treatment is new or is not widely used (as is often the case in drug trials), then the choice for π_C is usually based on past experience of the disease in the population to be studied.

Section 11 Choosing the sample size

The easiest way to decide on a value for π_T is usually to begin by specifying the difference $\pi_T - \pi_C$. This difference should correspond to the treatment effect that the trial is required to detect. This in turn will often be determined by what constitutes a difference that is relevant or useful in practice. This is sometimes called a **practically significant** difference. The term 'significant' here refers solely to the size of the effect. *Practical* significance differs from *statistical* significance, which involves probabilities. For example, a very small difference of no practical relevance could be statistically significant in a very powerful trial. Conversely, a practically significant effect may not turn out to be statistically significant if the power of the trial is too low.

Example 11.5 Whooping cough vaccine trial

In 1974, serious concerns arose over the safety of the whooping cough vaccine, following much publicized reports of brain damage following vaccination. Although the vaccine was later shown to be safe with respect to serious events such as brain damage, it is known to cause mild reactions such as redness and swelling at the injection site. A new vaccine was developed in the hope that such reactions could be reduced. This new vaccine was tested against the existing vaccine in several Phase II trials.

Suppose that it is required to design a trial with two groups to test the hypothesis that the new vaccine produces fewer reactions at the injection site than the standard vaccine. Children allocated to the treatment group (T) will receive the new vaccine. Children allocated to the control group (C) will receive the standard vaccine. The trial hypotheses are

$$H_0 : p_T = p_C, \quad H_1 : p_T \neq p_C,$$

where p_T and p_C are the underlying proportions of children experiencing reactions at the injection site among children given the new vaccine and among children given the existing vaccine. The significance level is chosen to be 0.05, and the power required is 0.8. The proportion of children who experience reactions at the injection site with the standard vaccine is known to be about 0.3, and discussions with doctors suggest that they would regard a reduction of 0.1 as practically significant, so that $\pi_T - \pi_C = -0.1$. Hence the design values are chosen to be $\pi_C = 0.3$ and $\pi_T = 0.3 - 0.1 = 0.2$. Hence $\pi_0 = 0.25$ and, using Formula (11.1), the sample size required in each group is given by

$$\begin{aligned} n &= \frac{2(q_{1-\alpha/2} + q_\gamma)^2 \pi_0(1-\pi_0)}{(\pi_T - \pi_C)^2} \\ &= \frac{2 \times (1.96 + 0.8416)^2 \times 0.25 \times (1 - 0.25)}{(-0.1)^2} \\ &= 294.33\ldots \\ &\simeq 295. \end{aligned}$$

Thus, in total, the number of children required is $2 \times 295 = 590$. ♦

In many circumstances, there is uncertainty about what value of π_C to use. In Example 11.5, for instance, the proportion of children with reactions using the standard vaccine may not be known exactly: the evidence might suggest that it probably lies between 0.2 and 0.4. In this case, it is a good idea to calculate sample sizes for a range of values to see what the impact of using the wrong value may be. This procedure is called a **sensitivity analysis**.

Activity 11.4 Sensitivity analysis for the whooping cough vaccine trial size

In the whooping cough vaccine trial of Example 11.5, a total sample size of 590 was required when $\pi_C = 0.3$. Suppose now that the evidence suggests that the value of π_C probably lies between 0.2 and 0.4.

(a) Using a significance level of 5%, power 0.8, and a practically significant difference of -0.1 as in Example 11.5, calculate the total sample size required for each of the values $\pi_C = 0.2$ and $\pi_C = 0.4$.

(b) What total sample size would you recommend to allow for the uncertainty in π_C?

In some circumstances there may be a practical limit on the trial size — for example, owing to budget constraints or limits on the numbers of participants available. In this case, the sample size cannot exceed a specified maximum. If the calculated sample size exceeds this maximum, it may be worthwhile examining what would be the power of a trial designed with the maximum sample size available. To obtain the power γ, the quantile q_γ must first be obtained by rearranging the expression for the sample size given in (11.1). An expression for q_γ is given in the following box.

Calculating the power for a given sample size

The power γ available in a trial with two groups each of size n is obtained from q_γ, the γ-quantile of $N(0,1)$, which is given by the following expression:

$$q_\gamma = |\pi_T - \pi_C| \sqrt{\frac{n}{2\pi_0(1-\pi_0)}} - q_{1-\alpha/2}. \quad (11.2)$$

In this expression, α is the significance level, π_T and π_C are the design values, $\pi_0 = (\pi_T + \pi_C)/2$, and $q_{1-\alpha/2}$ is the $(1-\alpha/2)$-quantile of $N(0,1)$.

The term $|\pi_T - \pi_C|$ in (11.2) is the absolute value (or magnitude) of the difference $\pi_T - \pi_C$.

Calculating the power for a given sample size is illustrated in Example 11.6.

Example 11.6 Power in the whooping cough vaccine trial

Suppose that budget constraints on the whooping cough vaccine trial described in Example 11.5 impose a limit of 500 children on the total trial size, that is, 250 in each group. For a significance level of 5% and design values $\pi_T = 0.2$ and $\pi_C = 0.3$ (as in Example 11.5), the value of q_γ is obtained using Formula (11.2):

$$q_\gamma = |\pi_T - \pi_C| \sqrt{\frac{n}{2\pi_0(1-\pi_0)}} - q_{1-\alpha/2}$$

$$= |0.2 - 0.3| \times \sqrt{\frac{250}{2 \times 0.25 \times (1-0.25)}} - 1.96$$

$$\simeq 2.582 - 1.96$$

$$\simeq 0.62.$$

Reference to the table of probabilities for the standard normal distribution in the *Handbook* gives $\gamma = 0.7324 \simeq 0.73$. Thus the achievable power with this sample size is 73%. ♦

See Table 1 in the *Handbook*.

Section 11 Choosing the sample size

Randomized controlled trials are usually designed to investigate one scientific hypothesis of primary importance, and the sample size is chosen for this purpose. However, it is common for other outcomes to be studied as well: these are known as **secondary outcomes** to distinguish them from the **primary outcome** which provides the main focus and justification for the trial. The power available to investigate such secondary outcomes is thus of interest. Activity 11.5 gives an example of such a situation, and also provides further practice with the material in this section.

Activity 11.5 Primary and secondary outcomes

In Activity 11.1, a trial of Infliximab, a new drug to treat rheumatoid arthritis, was discussed. You were asked to specify the null and alternative hypotheses for a trial to test the scientific hypothesis that Infliximab increases the proportion of patients who undergo a 50% improvement in their condition within one year (the ACR50).

(a) Obtain the sample size required in each group, with significance level 0.05, power 0.9, $\pi_C = 0.25$ (where π_C is the ACR50 for the control group), and practically significant difference $\pi_T - \pi_C = 0.1$.

(b) In addition to the primary outcome, it is of interest to compare the proportions of patients in the two groups that experience a severe adverse event (in case the treatment causes severe side effects). Using the sample size you obtained in part (a) and significance level 0.05, calculate the power available to detect an increase of 5% in the proportion of patients with severe adverse events, if the underlying proportion of patients on the standard therapy plus placebo with such events is 10%.

Summary of Section 11

In this section, you have seen that, before the sample size for a randomized controlled trial can be calculated, a clear statement is required of the trial hypotheses. The significance level and power must also be specified, together with the design values π_T and π_C. The value of π_T may be obtained by first identifying the practically significant difference $\pi_T - \pi_C$. An expression has been given for the size of each group. This expression depends on the significance level to be used, the power required and the design values. You have used this expression directly to calculate sample sizes. And you have used a formula obtained by rearranging this expression to calculate the power available for secondary comparisons for a fixed group size.

Exercise on Section 11

Exercise 11.1 Sample size for the pneumococcal vaccine trial

This exercise is based on the pneumococcal vaccine trial described in Exercises 10.1 and 10.2. The sample size required for the trial was calculated using significance level 0.05, power 0.8, and the design values $\pi_C = 0.55$ (the proportion of children in the control group with at least one episode of otitis media) and $\pi_T = 0.40$ (the proportion of children in the treatment group with at least one episode of otitis media).

(a) Based on these values, calculate the sample size required per group.

(b) It was anticipated that there would be 10% losses to follow-up. Calculate the total number of children that should be randomized.

12 Combining evidence from several studies

Randomized controlled trials are often regarded as the best means of providing reliable evidence of an association between a treatment or intervention and a specified disease outcome. However, even RCTs can fail to provide unambiguous evidence of causation. For example, they may lack power, or trials undertaken in different populations might produce different results. Thus, in order to draw definitive conclusions about causality and the likely impact of an intervention, reliance is seldom placed on a single study: the evidence from all trials, cohort studies and case-control studies is examined.

The idea of looking at all the evidence, not just that from a single trial or study, is in line with Bradford Hill's third criterion for causality, namely that of *consistency*: the association between an exposure E and a disease D must be apparent in several independent studies.

Bradford Hill's criteria were described in Subsection 8.1.

How is consistency of an association across studies assessed? The basic idea is very simple: assemble data from most or all of the studies that have been undertaken to investigate the association between E and D, and compare the results. If all such studies show a positive association, say, then the evidence for a causal association will be stronger than that obtained from each separate study.

Such reviews of all the available data have become an important feature of modern medical statistics. They have been greatly facilitated by contact and collaboration between research teams, availability of central reference databases, and electronic means of communication. New qualitative and quantitative statistical methods have been developed to improve the quality of such reviews.

The key aspect of these reviews is that they are *systematic*, whence the term **systematic review** is used to describe them. A systematic review may include only the randomized controlled trials undertaken on the topic, or it may include all studies including trials and epidemiological studies. In this section, you will meet examples of both.

In Subsection 12.1, systematic reviews are discussed briefly. The Mantel–Haenszel method is used to combine the results from several studies to obtain a pooled estimate of the odds ratio. A graphical method of summarizing the information from a group of studies is described in Subsection 12.2.

12.1 Systematic reviews and meta-analysis

A systematic review is conducted in the same way as a good randomized controlled trial or epidemiological study, with the key difference that the units are studies rather than persons. A study plan is set out, with clear criteria about which studies to include and which not to include. Much effort is devoted to obtain access to most or all studies in the particular area of interest, using databases such as Medline or the Cochrane registers. The selection procedure is important: a haphazard selection of only the better-known studies is likely to introduce a selection bias, since such studies are more likely to have yielded significant results. It can also be important to try to gain access to unpublished data sets as well as published data sets in order to reduce **publication bias**. This bias arises because authors may be more inclined to submit for publication, and journal editors more inclined to accept, research findings that show 'statistically significant' effects.

Once the search strategy and inclusion criteria have been decided, checklists are drawn up to score studies against reliability and quality criteria. These include, for example, the study type (randomized controlled trials, cohort studies, case-control studies) and the likely sources of bias and confounding, in order to evaluate the relative merits of the studies.

In some cases, it is possible to supplement a primarily *qualitative* systematic review with a *quantitative* assessment, by combining the numerical results from several studies to obtain a single summary estimate. This type of quantitative analysis of several different studies is called **meta-analysis**. One purpose of a meta-analysis is to obtain a better estimate of the strength of association between the exposure E and the disease D, by pooling the results of several studies.

The idea of combining information across studies dates back at least to Karl Pearson, long before the name meta-analysis was invented.

Example 12.1 Systematic review of the effectiveness of traffic calming measures

At the beginning of the 21st century, it was estimated that, throughout the world, over a million people died each year from traffic injuries and ten million people sustained permanent disabilities. In urban areas of high-income countries, area-wide traffic calming schemes have been proposed as a strategy to reduce accidents and casualties.

In a systematic review of the impact of such schemes, it was decided to investigate all randomized controlled trials and non-randomized controlled studies of such schemes involving certain specific changes to the road layout (road narrowing, speed humps, road closures, road surface treatment, changes at junctions, etc.). The studies were controlled in that they included study periods both before and after the changes were introduced.

Bunn, F., Collier, T., Frost, C., Ker, K., Roberts, I. and Wentz, R. (2003) Traffic calming for the prevention of road traffic injuries: systematic review and meta-analysis. *Injury Prevention*, **9**, 200–204.

Eleven relevant databases were searched. From these, 12 986 published and unpublished reports on traffic calming were identified and screened for eligibility. Twelve reports describing sixteen studies meeting the inclusion criteria were selected and included in the review. Of these sixteen studies, eleven found that road traffic injuries were less frequent after traffic calming was introduced, and five found that they were more frequent.

The authors combined the results of the sixteen studies in a meta-analysis. Overall, they concluded that the relative risk of fatal and non-fatal road traffic injuries after traffic calming compared to before traffic calming was 0.89, with 95% confidence interval (0.80, 1.00). Thus traffic calming measures reduced the accident rate by an estimated 11%. ♦

How should the information from several studies be combined? There are several approaches to this, but in this subsection just one method will be considered.

Suppose that k independent studies of the same type (cohort or case-control) have been identified through a systematic review of all studies to evaluate the association between an exposure E and a disease D. One way to think about the individual studies is as *strata* within the meta-analysis. Thus the estimated odds ratios $\widehat{OR}_1, \ldots, \widehat{OR}_k$ for the k studies can be combined using the Mantel–Haenszel method, which was described in Subsection 7.1, to give a pooled estimate of the odds ratio, together with a confidence interval.

A similar idea can be applied to pooling relative risks and other effect measures, but these will not be considered here.

A very important issue in meta-analysis is the extent to which the odds ratios vary between studies, and whether such variation is consistent with the underlying odds ratio being constant. Tarone's test for homogeneity (which was described in Subsection 7.3) can be used to test the null hypothesis $OR_1 = OR_2 = \ldots = OR_k$. If the null hypothesis is rejected, the studies are said to be **heterogeneous**.

The use of the Mantel–Haenszel method to obtain a pooled estimate of the odds ratio and Tarone's test for homogeneity in meta-analysis are illustrated in Example 12.2.

Example 12.2 Oral contraceptives and breast cancer

'*Pill users face 10-year tumour risk*' ran the headline in the *Sunday Times*. The year was 1996, and the *Lancet* was about to publish a large meta-analysis on the association between use of oral contraceptives ('the Pill') and breast cancer. As is often the case, particularly with health-scare stories, the attention-grabbing headline rather exaggerated the risk.

The study authors collected data from 54 epidemiological studies and undertook numerous analyses of these data. Table 12.1 shows data from one group of case-control studies known as nested studies (because the data form part of larger studies). In these studies, a woman is classified as exposed if she ever took oral contraceptives. If she never took any, she is classified as not exposed. Thus the exposed and not exposed groups are labelled 'Ever' and 'Never'.

Collaborative Group on Hormonal Factors in Breast Cancer (1996) Breast cancer and hormonal contraceptives: collaborative reanalysis of individual data on 53 297 women with breast cancer and 100 239 women without breast cancer from 54 epidemiological studies. *Lancet*, **347**, 1713–1727.

Table 12.1 Oral contraceptives and breast cancer: nested case-control studies

Study	Ever		Never		Odds ratio	95% CI
	Cases	Controls	Cases	Controls		
1	198	728	128	576	1.22	(0.96, 1.57)
2	96	437	101	342	0.74	(0.54, 1.02)
3	1105	4243	1645	6703	1.06	(0.97, 1.16)
4	741	2905	594	2418	1.04	(0.92, 1.17)
5	264	1091	907	3671	0.98	(0.84, 1.14)
6	105	408	348	1248	0.92	(0.72, 1.18)
7	138	431	436	1576	1.16	(0.93, 1.44)

None of the studies listed in Table 12.1 provided much evidence of an association between use of oral contraceptives and breast cancer. For example, study 1 included a total of 1630 cases and controls, and the estimated odds ratio was

$$\widehat{OR} = \frac{198 \times 576}{728 \times 128} \simeq 1.22.$$

The Mantel–Haenszel odds ratio is $\widehat{OR}_{MH} \simeq 1.04$. This is the pooled estimate of the underlying odds ratio obtained from these seven studies. The 95% confidence interval for the odds ratio is $(0.98, 1.10)$.

The observed value of the test statistic for Tarone's test for homogeneity is 8.72. The null distribution of the test statistic is $\chi^2(6)$, and the p value is 0.190. This indicates that there is little evidence of heterogeneity between studies.

The conclusion from this small subgroup of nested case-control studies is that there is little evidence of an association between ever having used oral contraceptives and breast cancer.

The pooled estimate of the odds ratio from all 54 studies reviewed was 1.07, with 95% confidence interval $(1.04, 1.10)$. The overall conclusion is that there is strong evidence of an increased risk of breast cancer from ever having used the Pill, but that the association is weak (that is, the odds ratio is close to 1). Further analysis showed that the risk is highest for current and recent Pill users, but that there is no evidence of an increased risk ten years after ceasing to use the Pill. ♦

Calculation of the pooled estimate of the odds ratio using the Mantel–Haenszel method is described in Subsection 7.1. Calculation of 95% confidence intervals and Tarone's test for homogeneity using SPSS are described in *Computer Book 1*.

12.2 Forest plots

A meta-analysis does not just consist of a single numerical summary. Arguably, the purpose of such an analysis is the presentation in succinct form of the data from several different studies: the differences between studies are at least as important as the overall pooled estimate. For large meta-analyses, presenting the raw data as in Table 12.1 may not be very informative. Instead, a special plot is used called a **forest plot**.

Example 12.3 *Forest plot for oral contraceptives and breast cancer studies*

The diagram in Figure 12.1 is a forest plot for the data in Table 12.1.

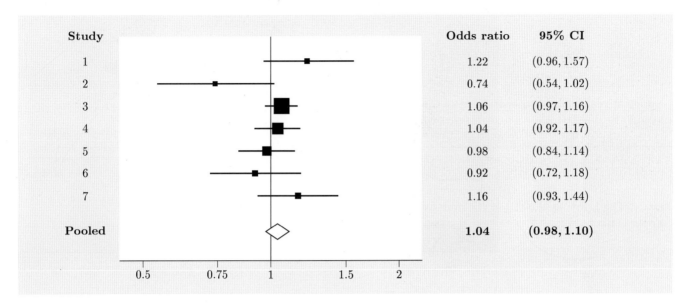

Figure 12.1 A forest plot for seven studies of breast cancer in ever-users compared to never-users of oral contraceptives

For each study, the study number or identifier is shown on the far left of Figure 12.1, and the odds ratio and 95% confidence interval are on the far right.

The distinctive feature of the plot is the arrangement of short horizontal lines around a central vertical spine. For each study, the value of the odds ratio is indicated by the position of a square, and the 95% confidence interval as a horizontal line segment. The central vertical line represents the odds ratio $OR = 1$. Thus a square to the right of this line corresponds to a study supportive of a positive association, a square to the left corresponds to a study supportive of a negative association.

The results for the seven studies combined are given below the results for the individual studies. These are labelled 'Pooled' in the left-hand column. The Mantel–Haenszel estimate of the odds ratio and its 95% confidence interval are given on the right-hand side. These are represented on the diagram by a diamond centered on the pooled estimate of the odds ratio and extending horizontally across the width of the 95% confidence interval.

The scale at the bottom of the plot in Figure 12.1 is a logarithmic scale. This means that odds ratios of 2 and $\frac{1}{2}$ are the same distance from the vertical line centred on $OR = 1$, which corresponds to no association.

Finally, note the variation in size of the squares indicating the positions of the odds ratios for the individual studies. The Mantel–Haenszel pooled estimate is a weighted sum of the odds ratios for the individual studies:

$$\widehat{OR}_{MH} = \sum_{i=1}^{k} w_i \widehat{OR}_i,$$

where the Mantel–Haenszel weights w_i sum to 1. The area of square i is proportional to the weight w_i — the larger the weight, the bigger the area. The purpose of this refinement is to attract attention to the studies which contribute most to the pooled odds ratio. These tend to be the largest studies. If the central squares were not scaled in this way, the most influential studies would be the least noticed since their confidence intervals tend to be the narrowest. This would distort the visual impression conveyed by the diagram. ♦

Different authors use different conventions for presenting forest plots — for example, the relative risk rather than the odds ratio may be used, and methods other than the Mantel–Haenszel method may be used to obtain the pooled estimate and the weights. However, all forest plots can be interpreted in a similar fashion. You will not be expected to draw forest plots, only to interpret them. Activity 12.1 will give you some practice with this.

Activity 12.1 Steroid injections and osteoarthritis

Osteoarthritis is a disease affecting the joints, often appearing in elderly people. A common treatment for osteoarthritis of the knee is injection of steroids into the knee joint. In 2004 a meta-analysis was published examining the evidence for clinical improvement following such interventions.

The authors identified six randomized placebo-controlled trials meeting their inclusion criteria, and which had data on patient improvement following treatment. Figure 12.2 shows the forest plot for the meta-analysis of these six trials. The arrow at the end of the horizontal line for study 2 indicates that the upper confidence limit lies beyond the range of the plot.

Arroll, B. and Goodyear-Smith, F. (2004) Corticosteroid injections for osteoarthritis of the knee: meta-analysis. *British Medical Journal*, **328**, 869–870.

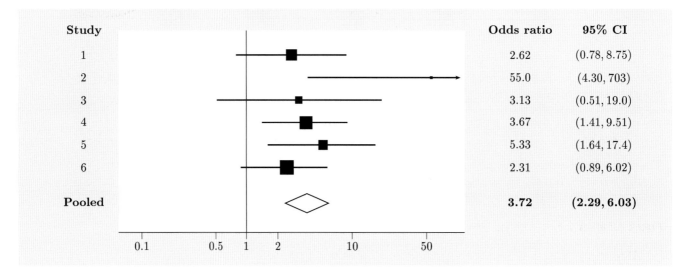

Figure 12.2 A forest plot for six randomized placebo-controlled trials of steroid injections and osteoarthritis of the knee

(a) Write down the pooled estimate of the odds ratio and its 95% confidence interval. What do you conclude from this? How does the forest plot convey this conclusion?

(b) The observed value of the test statistic for Tarone's test for homogeneity was 6.76. What do you conclude?

(c) Which study contributed most to the pooled estimate of the odds ratio? Which contributed least? How is this represented on the forest plot?

Summary of Section 12

In this section, methods for comparing the results of different trials and other epidemiological studies have been described. Systematic reviews provide a framework for evaluating evidence from several studies. In some cases it is possible to undertake a meta-analysis to obtain a pooled estimate of the odds ratio. The information from the studies in a meta-analysis may be summarized in a forest plot. You have learned how to interpret forest plots.

Exercise on Section 12

Exercise 12.1 Breast cancer and HRT

In Example 2.2, the Million Women Study of the association between invasive breast cancer and hormone-replacement therapy (HRT) was described. This study was an uncommonly large cohort study, undertaken to obtain a clear answer to the question 'Does HRT cause breast cancer?'.

Clearly, one does not embark on such a large study lightly. Before it was undertaken, many other smaller studies had been done. In 1997, a meta-analysis was published of 51 studies of HRT and breast cancer, undertaken in 21 countries.

The forest plot in Figure 12.3 is based on a subset of the data, corresponding to six case-control studies with controls sampled from the same hospitals as the cases. Exposure is defined as ever having used HRT. Women who have never used HRT are unexposed.

Collaborative group on Hormonal Factors in Breast Cancer (1997) Breast cancer and hormone replacement therapy: collaborative reanalysis of data from 51 epidemiological studies of 52 705 women with breast cancer and 108 411 women without breast cancer. *Lancet*, **350**, 1047–1059.

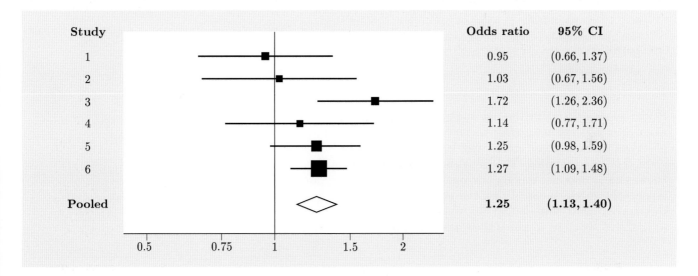

Figure 12.3 A forest plot for six case-control studies of breast cancer and HRT

(a) How many of the studies support a positive association? How many support a negative (that is, protective) association?

(b) The observed value of the test statistic for Tarone's test for homogeneity was 7.31. What do you conclude?

(c) What overall conclusion do you draw about the association, or otherwise, between HRT and breast cancer?

13 The medical literature

Medical statisticians are often called upon to read, assess and comment on reports of epidemiological studies and randomized controlled trials. In this section, you are invited to read a published article, and apply the knowledge you have acquired in this book to interpret the article.

Increasingly, medical journals make use of the facilities offered by the internet. Online publication means that a paper appears earlier than its print version and is more widely available. The *British Medical Journal* is no exception. Moreover, the articles published in the print version of the journal are often abridged versions of full papers published online. The article you will read is the full version of a paper, referenced as follows.

> Gertsch, J.H., Basnyat, B., Johnson, E.W., Onopa, J. and Holck, P.S. (2004) Randomised, double blind, placebo controlled comparison of ginkgo biloba and acetazolamide for prevention of acute mountain sickness among Himalayan trekkers: the prevention of high altitude illness trial (PHAIT). *British Medical Journal*, doi:10.1136/bmj.38043.501690.7C (first published 11 March 2004).

You may find it useful to obtain a further copy of the paper to work through. An electronic version of the paper is available on the *M249* website.

The paper has been reprinted in Subsection 13.1.

Before you start working through the paper, it is important to point out that research papers are not perfect documents. They quite commonly contain misprints and arithmetic errors, and, less frequently, more fundamental errors of method or interpretation. This paper is no exception. Although, to our knowledge, it does not contain any fundamental errors, there are a few typographical errors, obscurities, and mistakes. You should read the paper with two aims in mind: first, to find out what the authors have done and what it means, and secondly, to assess their methods and findings critically.

13.1 The high altitude mountain sickness trial

The following article has been reproduced with permission from the *British Medical Journal*.

Like many reports of statistical analyses, the article comprises five main parts: Abstract, Introduction, Methods, Results, and Discussion, and is followed by a list of references (which you may ignore). Towards the end of the article is a box with two panels, entitled 'What is already known on this topic' and 'What this study adds'. This contains a very brief, non-technical account of the findings of the study. Locate this box and read it now to get a general idea of what the article is about. *Do not attempt to read the rest of the article yet.* You will work through the article step by step in Subsections 13.2 to 13.5. When you have read the box, go to Subsection 13.2.

Randomised, double blind, placebo controlled comparison of ginkgo biloba and acetazolamide for prevention of acute mountain sickness among Himalayan trekkers: the prevention of high altitude illness trial (PHAIT)

Jeffrey H Gertsch, Buddha Basnyat, E William Johnson, Janet Onopa, Peter S Holck on behalf of the Prevention of High Altitude Illness Trial Research Group

Abstract

Objective To evaluate the efficacy of ginkgo biloba, acetazolamide, and their combination as prophylaxis against acute mountain sickness.
Design Prospective, double blind, randomised, placebo controlled trial.
Setting Approach to Mount Everest base camp in the Nepal Himalayas at 4280 m or 4358 m and study end point at 4928 m during October and November 2002.
Participants 614 healthy western trekkers (487 completed the trial) assigned to receive ginkgo, acetazolamide, combined acetazolamide and ginkgo, or placebo, initially taking at least three or four doses before continued ascent.
Main outcome measures Incidence measured by Lake Louise acute mountain sickness score ≥ 3 with headache and one other symptom. Secondary outcome measures included blood oxygen content, severity of syndrome (Lake Louise scores ≥ 5), incidence of headache, and severity of headache.
Results Ginkgo was not significantly different from placebo for any outcome; however participants in the acetazolamide group showed significant levels of protection. The incidence of acute mountain sickness was 34% for placebo, 12% for acetazolamide (odds ratio 3.76, 95% confidence interval 1.91 to 7.39, number needed to treat 4), 35% for ginkgo (0.95, 0.56 to 1.62), and 14% for combined ginkgo and acetazolamide (3.04, 1.62 to 5.69). The proportion of patients with increased severity of acute mountain sickness was 18% for placebo, 3% for acetazoalmide (6.46, 2.15 to 19.40, number needed to treat 7), 18% for ginkgo (1, 0.52 to 1.90), and 7% for combined ginkgo and acetazolamide (2.95, 1.30 to 6.70).
Conclusions When compared with placebo, ginkgo is not effective at preventing acute mountain sickness. Acetazolamide 250 mg twice daily afforded robust protection against symptoms of acute mountain sickness.

Introduction

Acute mountain sickness is a syndrome that occurs above 2000 m secondary to failed physiological adaptation to acute hypobaric hypoxia. This rapidly reversible condition is characterised by headache, lightheadedness, fatigue, nausea, and insomnia. If untreated the condition may progress to life threatening high altitude cerebral oedema or pulmonary oedema.[1][2] Although modifiable aspects of high altitude travel such as ascent rate and exertion are the primary mediators of risk, pharmaceutical prevention with acetazolamide is also effective despite common side effects such as parasthesias, dysgeusia, and diuresis, which may reduce compliance. Acute mountain sickness is a common diagnosis at high altitude, and effective, readily available, and safer prophylactic agents are needed.

Ginkgo biloba is a popular herbal supplement, which has emerged as a new prophylactic agent for the prevention of acute mountain sickness.[3-8] Indirect evidence suggests that it may prevent hypoxic damage in tissues in part as a result of its antioxidant activity, and in clinical trials its side effects profile was similar to placebo.[9][10] Our group has shown that prophylactic ginkgo may lead to a reduction in acute mountain sickness, with no recognisable side effects, indicating that it may be a viable alternative to acetazolamide.[4] The results of multiple small randomised controlled trials with ginkgo have, however, been mixed.

To date there have been no large scale, randomised controlled trials comparing ginkgo with acetazolamide on prevention of acute mountain sickness or testing the two combined for safety and additive efficacy. We compared the effect either ginkgo, acetazolamide, or combined ginkgo and acetazolamide with placebo on the incidence and severity of acute mountain sickness and headache in people who trek at high altitudes.

Methods

Our study was designed as a prospective, randomised, double blind, placebo controlled trial. Enrolment took place between 6 October and 24 November 2002 along the Mount Everest approach in the Nepal Himalayas.

The predetermined primary outcome measure was incidence and severity of acute mountain sickness at the study end point as judged by the Lake Louise scoring system, a well validated standard for evaluation of acute mountain sickness in the field.[11-13] Acute mountain sickness is quantified on the Lake Louise questionnaire in a high altitude setting as a score of three or greater, with headache and at least one of the symptoms of nausea or vomiting, fatigue, dizziness, or difficulty sleeping. Predetermined secondary end points included incidence and severity of headache and endpoint pulse oximetry (Nonin Medical Products, Minneapolis, USA). Personal data, ascent profile, compliance, and side effects were analysed to discount potential confounders.

Our trial was double blinded, and the randomisation code was computer generated by Deurali-Janta Pharmaceuticals (Kathmandu, Nepal) and held by an independent physician. The standardised ginkgo extract GK 501 was manufactured by Pharmaton (Lugano, Switzerland) in strict accordance with German European Commission standards, with no less than 24% ginkgo-flavone glycosides and 6% terpenes. The acetazolamide was manufactured by Wyeth (Madison, USA). Samples from the randomised batch of study drugs were verified for purity and activity by Boehringer-Ingelheim (Germany).

Randomisation and follow up

Trekkers completed questionnaires after giving signed informed consent. Inclusion criteria specified healthy non-Nepali males and females aged 18-65 years travelling directly between the baseline villages of Pheriche or Dingboche (4280 m and 4358 m, respectively) and the end point in Lobuje (4928 m). Potential participants were excluded if they had acute mountain sickness, had signs and symptoms of a substantial acute infection, had slept above 4500 m, had taken ginkgo or acetazolamide within two weeks before enrolment, or had any known cardiac, pulmonary, or other chronic disease that would render them at increased risk of altitude illness.

Trekkers newly arrived at the baseline altitude were screened daily and serially enrolled by randomisation number. They completed the Lake Louise questionnaire, had pulse oximetry readings taken, and provided data on personal characteristics and rate of ascent. They were given information on methods for reducing the risk of acute mountain sickness. Participants were randomised in a double blind fashion to receive twice daily either ginkgo 120 mg, acetazolamide 250 mg, combined ginkgo 120 mg and acetazolamide 250 mg, or placebo. Participants took a minimum of three or four doses of the study drugs at baseline altitude before proceeding on their trek without any influence from study administrators. On their ascent from baseline, some participants stopped overnight at a lodge at 4595 m, but all were expected to arrive at the end point altitude for data collection (Lake Louise questionnaire, pulse oximetry, rate of ascent, and side effects). Lake Louise scores were taken the morning after arrival, after which the study was complete.

Power calculation and statistical analysis

Preliminary estimates suggested that we would need a minimum sample size of 70 people per group to determine a statistically significant difference (80% power) between treatment and placebo groups based on a published acute mountain sickness attack rate of 57% at the end point.[14] We used odds ratios and associated confidence intervals (asymptotic or Fisher's exact test) to estimate the effects of categorical variables (confidence intervals excluding one represent significant effects). Means of continuous outcomes were compared using t tests, and we considered P values less than 0.05 as significant.

Results

Overall, 614 trekkers were enrolled and 487 completed the study (figure). Participants in all four groups, including those lost to follow up, were similar at baseline (table 1). The 127 participants (20.7%) lost to follow up had similar personal characteristics to all participants (data not shown).

Table 2 summarises the main outcome profile for the 487 participants who completed the study. The data are presented as an intention to treat analysis; there were no significant changes

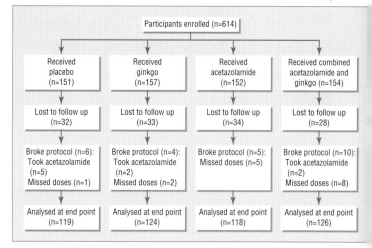

Flow of participants through trial

noted when the table was reproduced without the data from non-compliant participants (data not shown). Analysis for the primary end point showed that ginkgo did not reduce the incidence of acute mountain sickness when compared with placebo; ginkgo also failed to show a benefit in secondary analyses. The small increase in incidence and severity of acute mountain sickness or headache in the ginkgo group was not significant when compared with placebo. We found no significant adverse events in any group (aggressive allergic reactions or high altitude cerebral oedema or pulmonary oedema).

Acetazolamide as a control intervention was associated with a substantial decrease in acute mountain sickness and incidence and severity of headache, as well as improved blood oxygen desaturation with ascent when compared with placebo. In the acetazolamide group the numbers needed to treat were 4 to prevent one instance of acute mountain sickness, 7 to prevent one instance of severe acute mountain sickness, 3 to prevent one instance of headache, and 8 to prevent headaches of greater severity. When compared with acetazolamide the combined drug caused a marginally significant increase in the incidence of headache (odds ratio 1.82, 95% confidence interval 1.0 to 3.3, number needed to harm 9) but did not significantly affect other surrogate markers (incidence of acute mountain sickness 1.24, 0.6 to 2.6; severity of acute mountain sickness 2.19, 0.7 to 7.3; severity of headache 1.90, 0.3 to 10.6).

The groups were compared for inequality in several measures, but we found no bias or inequality at the end point excepting the typical side effects of acetazolamide (table 3).

Discussion

Ginkgo was not effective in reducing the incidence or severity of acute mountain sickness when compared with placebo and failed to show a protective benefit for any outcome measure. Furthermore, the addition of ginkgo to acetazolamide caused a marginally significant decrease in the efficacy of acetazolamide against headache (the most common symptom at altitude); this was unexpected considering the different proposed mechanisms of action for the two substances. Research has shown ginkgo to have some vasodilatory properties.[15] This may theoretically increase cerebral blood flow, which in turn could worsen the symptoms of acute mountain sickness such as headache. Regardless of the mechanism, clinicians should avoid recommending ginkgo as prophylaxis for acute mountain sickness either alone or combined with acetazolamide.

Table 1 Baseline characteristics of 487 of 614 participants who completed study on effects of prophylactic agents against acute mountain sickness

		Study groups			
Variables	All participants (n=487)	Placebo (n=119)	Acetazolamide group (n=118)	Ginkgo group (n=124)	Combined acetazolamide and ginkgo (n=126)
No (%) male	337 (69)	88 (74)	79 (67)	83 (67)	88 (70)
Mean (SD) age (years)	36.6 (10.9)	36.4 (10.8)	36.4 (11.0)	36.7 (10.5)	36.7 (11.4)
No (%) if trekkers starting from 2800 m*	395 (81)	91 (76)	101 (86)	104 (84)	99 (79)
Mean No (SD) nights to ascend from 2800 m	4.7 (1.3)	4.6 (1.2)	4.7 (1.1)	4.5 (1.9)	4.9 (1.5)
No (%) enrolled at 4358 m†	259 (53)	67 (56)	58 (49)	68 (55)	66 (52)
No (%) with baseline Lake Lousie score of 1‡	27 (5)	6 (5)	7 (6)	6 (5)	8 (6)
Mean (SD) baseline oxygen saturation	85.4 (4.3)	85.9 (4.3)	85.3 (4.4)	84.8 (4.8)	85.5 (3.7)
No (%) lost to follow up§	127 (26)	32 (27)	34 (29)	33 (27)	28 (22)

*Lukla airport is at about 2800 m, and trekkers starting from Jiri (2000 m) pass through Lukla.
†Enrolment occurred in villages of Dingboche (4358 m) and Pheriche (4280 m).
‡Most participants at baseline scored zero.
§From original 614 participants.

This is the first study in which ginkgo prophylaxis was given when the participants were enrolled at a high baseline altitude (as opposed to starting the drug at sea level before ascent). This may explain why our results were negative compared with previous trials. Other reasons include the quality and purity of the ginkgo preparations, the dose of ginkgo used, the number of days in which ginkgo was preloaded before controlled ascent, and environmental or behavioural influences on ginkgo's effectiveness.

Our study is among the largest randomised trials of acetazolamide for acute mountain sickness prophylaxis, and the 250 mg twice daily regimen exhibited a robust and predictable

Table 2 Main outcome profile (intent to treat) in groups treated with prophylactic agents for acute mountain sickness. Values are numbers (percentages) unless stated otherwise

Variables	All participants (n=487)	Placebo group (n=119)	Acetazolamide group (n=118)	Odds ratio (95% CI)*	Ginkgo group (n=124)	Odds ratio (95% CI)*	Combined acetazolamide and ginkgo group (n=126)	Odds ratio (95% CI)*
Incidence of acute mountain sickness†	115 (24)	40 (34)	14 (12)	3.76 (1.91 to 7.39)	43 (35)	0.95 (0.56 to 1.62)	18 (14)	3.04 (1.62 to 5.69)
Severe acute mountain sickness‡	58 (12)	22 (18)	4 (3)	6.46 (2.15 to 19.40)	23 (18)	1.00 (0.52 to 1.90)	9 (7)	2.95 (1.30 to 6.70)
Headache incidence	197 (40)	63 (53)	23 (19)	4.77 (2.70 to 8.44)	72 (58)	0.75 (0.47 to 1.22)	39 (31)	2.62 (1.58 to 4.35)
Severe headache§	46 (9)	16 (13)	2 (2)	9.01 (2.02 to 40.13)	24 (19)	0.65 (0.32 to 1.29)	4 (3)	4.74 (1.54 to 14.62)
Mean endpoint oxygen saturation (%)¶	82.3	82.1	83.7	P=0.01	79.5	P<0.01	83.9	P<0.01
Decrease in oxygen saturation from baseline¶	3.0	3.8	1.7	P<0.01	5.2	P=0.04	1.5	P<0.01
Non-compliant**	25 (5)	6 (5)	5 (4)	0.75 (0.75 to 2.10)	4 (3)	1.49 (0.46 to 4.76)	10 (8)	0.59 (0.22 to 1.56)

*Compared with incidence in placebo group.
†Lake Louise score ≥3 with headache and at least one other symptom.
‡Lake Louise score ≥5.
§Determined by cut off between scores of 1 and 2 on the Lake Louise survey (ascending scale of 0-3 for severity).
¶Mean values reported, with P values of t tests comparing differences in mean of treatment group to mean in placebo group.
**Acetazolamide taken outside study protocol or ≥3 consecutive study doses missed.

Table 3 Ascent rate, compliance, and side effects of groups receiving prophylactic agents for acute mountain sickness. Values are numbers (percentages) unless stated otherwise

Variables	All participants (n=487)	Placebo group (n=119)	Acetazolamide group	Odds ratio (95% CI)*	Ginkgo group (n=124)	Odds ratio (95% CI)*	Combined acetazolamide and ginkgo group (n=126)	Odds ratio (95% CI)*
Acclimatisation nights:								
Baseline†	348 (71)	85 (71)	81 (69)	1.14 (0.65 to 1.99)	93 (75.0)	0.83 (0.47 to 1.47)	89 (71)	1.04 (0.60 to 1.81)
Midpoint (4595 m)	95 (19)	24 (20)	26 (22)	0.89 (0.48 to 1.67)	14 (11.3)	1.98 (0.97 to 4.05)	31 (25)	0.77 (0.42 to 1.42)
Doses missed	33 (7)	7 (6)	9 (8)	1.16 (0.34 to 3.90)	5 (4.0)	1.59 (0.44 to 5.80)	12 (9)	0.76 (0.26 to 2.27)
Paraesthesias	200 (41)	12 (10)	85 (72)	0.04 (0.02 to 0.09)	10 (8.1)	1.28 (0.53 to 3.08)	93 (74)	0.04 (0.02 to 0.08)
Blurred vision	7 (1)	3 (2)	0 (0)	NA	2 (1.6)	0.64 (0.18 to 19.16)	2 (2)	0.62 (0.18 to 19.48)
Rash	7 (1)	1 (1)	0 (0)	NA	0 (0.0)	NA	6 (5)	0.17 (0.01 to 1.44)
Frequency	21 (4)	2 (2)	10 (8)	0.18 (0.04 to 0.86)	2 (1.6)	1.04 (0.14 to 7.52)	7 (6)	0.29 (0.06 to 1.43)
Dysgeusia	33 (7)	3 (2)	13 (11)	0.21 (0.06 to 0.75)	6 (4.8)	0.51 (0.12 to 2.08)	11 (9)	0.27 (0.07 to 0.99)

*Compared with incidence in placebo group.
†Two nights.

clinical effect closely in line with previous studies.[16 17] These results validate acetazolamide therapy as the standard of care for pharmacological prevention of acute mountain sickness, which may be used as an adjunct to behavioural strategies for avoiding altitude sickness. The substantial clinical effectiveness of the acetazolamide 250 mg twice daily regimen in this trial is also important in light of the group's previous results from a randomised trial of similar design, which utilised acetazolamide 125 mg twice daily and also showed significant protection against acute mountain sickness. The combined data from these prospective studies clearly counter the results of a meta-analysis published previously, which suggested that at least 750 mg of acetazolamide daily is required for adequate prophylaxis against acute mountain sickness.[17 18]

Limitations of the study
Our study had several limitations. Firstly, baseline was at a high elevation (4280 m or 4358 m); many participants will have acute mountain sickness below this altitude, and hence it is difficult to compare our results with those of other studies that had a low baseline altitude. Although it may be argued that participants who achieve ascent to this altitude are relatively resistant to acute mountain sickness, these individuals were protected from acute mountain sickness using acetazolamide in a manner that was consistent with previous studies, suggesting that testing conditions were adequate.[16 17] Secondly, just over a fifth of participants were lost to follow up, and the outcome among these people may have affected the significance of our findings. However, participants in all groups were equally likely to drop out of the study, and a reasonable degree of attrition is expected (and consistent with previous studies) owing to the wilderness setting and the low incentive to follow up at the study end point.[14 16 17] Thirdly, participants were recruited at two villages situated in close proximity, with around 78 m difference in elevation, which may differentially influence the degree of exposure to hypoxia. We did not, however, consider this important, as judged by the lack of any statistically significant differences in groups for personal characteristics or outcome measures (data not shown). Lastly, although we studied a diverse population in typical trekking conditions, these results may not be generalisable to other high altitude trekking environments where ascent rates and baseline elevation or final elevation may be different.

We thank Pharmaton of Lugano; Deurali-Janta Pharmaceuticals of Kathmandu, Nepal and Hari Bhakta Sharma for randomisation of the drugs and packaging; Eric Johnson for clinical support; Anna Donahue, Joel Meyer, and Sabina Yamamura for their translations of study materials into Italian and Spanish, French, and German, respectively; and the trekkers for their participation. Members of the Prevention of High Altitude Illness Trial (PHAIT) Research Group are Brendon Cogtry (Medical College of Georgia, USA), Amy Derrow (Wake Forest University School of Medicine, NC, USA), Danielle Douglas (University of California, San Diego, CA), Joel Meyer and Stephen G Seale (Royal United Hospital, Bath, UK), Allison Mulcahy (University of Nevada School of Medicine, Reno, USA), Jessica Ngo (Stanford School of Medicine, CA, USA), Christian Purgason (Arizona College of Osteopathic Medicine, Glendale, USA), Joanne Snow (Los Angeles Harbor Medical Center, Los Angeles, CA). and Hassan Zacharia (Medical College of Virginia, Richmond, VA).

Contributors: JHG wrote the original manuscript with the close consultation of the authors; he will act as guarantor for the paper. The guarantor accepts full responsibility for the conduct of the study, had access to the data, and controlled the decision to publish. All authors were responsible for the study design with end stage input from BM Cogtry, AE Derrow, DJ Douglas, JY Meyer, AR Mulcahy, JD Ngo, CT Purgason, SG Seale, JL Snow, and H Zacharia. JO and JHG acquired the funding. JHG, BB, BM Cogtry, AE Derrow, DJ Douglas, JY Meyer, AR Mulcahy, JD Ngo, CT Purgason, SG Seale, JL Snow, and H Zacharia were responsible for the implementation of the study and data collection. JHG, BB, EWJ, JO, and PSH were responsible for data entry and analysis. JHG, BB, EWJ, JO, PSH,

What is already known on this topic

Ginkgo biloba, an experimental prophylactic agent against acute mountain sickness, has shown mixed efficacy in several small randomised controlled trials

Acetazolamide is the standard pharmaceutical prevention of acute mountain sickness

The minimum effective dose of acetazolamide is under debate

What this study adds

This large randomised controlled clinical trial showed that ginkgo was not effective in decreasing the incidence or severity of acute mountain sickness

The efficacy of acetazolamide for preventing headache was decreased when combined with ginkgo

Acetazolamide at 500 mg daily had a robust clinical effect

Acetazolamide could be used as an adjunct to behavioural strategies for avoiding altitude sickness

BM Cogtry, AE Derrow, DJ Douglas, JY Meyer, AR Mulcahy, JD Ngo, CT Purgason, SG Seale, JL Snow, and H Zacharia prepared and revised the manuscript.

Funding: Pharmaton provided financial support for study expenses. Representatives of Pharmaton provided limited statistical support by generating the power calculation.

Competing interests: JHG and JO have been funded by Pharmaton to attend a research symposium. All authors except BB, EWJ, JO, and PSH have received reimbursement for on-site living costs incurred during the implementation period of the study.

Ethical approval: This study was conducted under the auspices of the Himalayan Rescue Association and received ethical approval from the Nepal Health Research Council.

1. Hackett PH, Roach RC. High altitude illness. *N Engl J Med* 2001;345:107-14.
2. Basnyat B. Altitude sickness. In: Rakel RE, Bope ET, eds. *Conn's current therapy*. Philadelphia: WB Saunders, 2001:1166-9.
3. Roncin JP, Schwartz F, D'Arbigny P. EGb 761 in control of acute mountain sickness and vascular reactivity to cold exposure. *Aviat Space Environ Med* 1996;67:445-52.
4. Gertsch JH, Seto TB, Mor J, Onopa J. Ginkgo biloba for the prevention of severe acute mountain sickness starting one day before rapid ascent. *High Alt Med Biol* 2002;2:110.
5. Leadbetter GW, Maakestad K, Olson S, Hackett PH. Ginkgo biloba reduces incidence and severity of acute mountain sickness. Abstracts from the 12th international hypoxia symposium, Jasper, Alberta, Canada, March 10-14, 2001. *High Alt Med Biol* 2001;2:110.
6. Chow TK, Browne VA, Heileson HL, Wallace DR, Anholm JD. Comparison of Ginkgo biloba versus acetazolamide in the prevention of acute mountain sickness. *Med Sci Sports Exerc* 2002;34:S246.
7. Leadbetter GW, Hackett PH, Maakestad K, Tissot-Van Patot M, Olson S, Keyes L, et al. Comparison of Ginkgo biloba, acetazolamide, and placebo for prevention of acute mountain sickness. Abstracts from the 13th international hypoxia symposium, Banff, Alberta, Canada, February 19-22, 2003. *High Alt Med Biol* 2003;3:455.
8. Moraga F, Flores A, Zapata J, Ramos P, Madariaga M, Serra J. Ginkgo biloba decreases acute mountain sickness (AMS) at 3700 m. Abstracts from the 13th international hypoxia symposium, Banff, Alberta, Canada, February 19-22, 2003. *High Alt Med Biol* 2003;3:453.
9. Le Bars PL, Katz MM, Berman N, Itil TM, Freedman AM, Schatzberg AF, for the North American EGb Study Group. A placebo-controlled, double-blind, randomized trial of an extract of Ginkgo biloba for dementia. *JAMA* 1997;278:1327-32.
10. Bailey DM, Davies B. Acute mountain sickness: prophylactic benefits of antioxidant vitamin supplementation at high altitudes. *High Alt Med Biol* 2001;2:21-9.
11. Roach RC, Bartsch P, Hackett PH, Oelz O. The Lake Louise AMS Scoring Consensus Committee. The Lake Louise acute mountain sickness scoring system. In: Sutton JR, Coates G, Huston CS, eds. *Hypoxia and molecular medicine: proceedings of the 8th international hypoxia symposium*, 9-13 Feb, 1993; Lake Louise, Alberta, Canada. Burlington, VT: Queen City Printers, 1993:272-4.
12. Bartsch P, Muller A, Hofstettler D, Maggiorini M, Vock P, Oelz O. AMS and HAPE scoring in the Alps. In: Sutton JR, Houston CS, Coates G, eds. *Hypoxia and molecular medicine: proceedings of the 8th international hypoxia symposium*, Lake Louise, Canada. Burlington, VT: Queen City printers, 1993:265-71.
13. Maggiorini M, Muller A, Hofstetter D, Bartsch P, Oelz O. Assessment of acute mountain sickness by different score protocols in the Swiss Alps. *Aviat Space Environ Med* 1998;69:1186-92.

14 Murdoch DR. Symptoms of infection and altitude illness among hikers in the Mount Everest region of Nepal. *Aviat Space Environ Med* 1995;66:148-51.
15 Chen X, Salwinski S, Lee TJ. Extracts of Ginkgo biloba and ginsenosides exert cerebral vasorelaxation via a nitric oxide pathway. *Clin Exp Pharmacol Physiol* 1997;24:958-9.
16 Hackett PH, Rennie D, Levine HD. The incidence, importance, and prophylaxis of acute mountain sickness. *Lancet* 1976;2:1149-54.
17 Basnyat B, Gertsch JH, Johnson EW, Castro-Marin F, Inoue Y, Yeh C. Efficacy of low-dose acetazolamide (125 mg BID) for the prophylaxis of acute mountain sickness. *High Alt Med Biol* 2003;4:45-52.
18 Dumont L, Mardirosoff C, Tramer MR. Efficacy and harm of pharmacological prevention of acute mountain sickness: quantitative systematic review. *BMJ* 2000;321:267-72.

(Accepted 23 January 2004)

doi 10.1136/bmj.38043.501690.7C

doi 10.1136/bmj.38043.501690.7C

Department of Internal Medicine, Maricopa Medical Center, 2601 E Roosevelt Avenue number O-D-10, Phoenix, AZ 85008, USA
Jeffrey H Gertsch *house officer*

Himalayan Rescue Association, Kathmandu, Nepal

Buddha Basnyat *medical director*

University of Washington School of Medicine, Seattle, WA 98195-6410, USA
E William Johnson *house officer*

University of Hawaii, John A Burns School of Medicine, Honolulu, Hawaii, USA
Janet Onopa *assistant professor of medicine*

Peter S Holck *associate professor of biostatistics*

Correspondence to: J H Gertsch
jeffgertsch@hotmail.com

> **Amendment**
>
> This is version 2 of the paper. In this version the point estimate for incidence of acute mountain sickness with combined acetazolamide and gingko has been corrected and now reads 1.24 [instead of 12.4] and the third footnote in table 2 now reads ‡Lake Louise score ≥ 5 ["&period" has been removed].

Gertsch, J.H. *et al.* 'Randomised, double blind, placebo controlled comparison of ginkgo biloba and acetazolamide for prevention of acute mountain sickness among Himalayan trekkers: the prevention of high altitude illness trial (PHAIT)'. *British Medical Journal*, 3 April 2004, Vol. 328, No. 7443.
© 2004 The BMJ Publishing Group

13.2 The Abstract and Introduction

The **Abstract** provides a succinct description of the aims, methods, results and conclusions of the study. The abstract to this article is in a very particular form called a **structured abstract**. This includes short paragraphs entitled *Objective, Design, Setting, Participants, Main outcome measures, Results,* and *Conclusions*. A structured abstract makes it possible to obtain an accurate overall impression of the study quickly.

Activity 13.1 Reading the Abstract

Read the Abstract now.

Comment

The word 'incidence' used in the paragraph entitled *Main outcome measures* refers to the proportion of trekkers with a particular syndrome. Reference is also made to the 'Lake Louise acute mountain sickness score': at this stage you need not worry about what this is — some clarification is provided in the article.

The *Results* paragraph includes odds ratios and confidence intervals, but also a new quantity, the number needed to treat. This is explained in the text following this activity.

After the first comparison, the reporting of subsequent odds ratios and confidence intervals is abbreviated. For example, the phrase '35% for ginkgo (0.95, 0.56 to 1.62)' means 'The incidence of acute mountain sickness was 35% in the ginkgo group, the odds ratio relative to placebo was 0.95, with 95% confidence interval (0.56, 1.62).'

The **number needed to treat** (NNT) is the average number of patients who must be treated in order to prevent one adverse disease outcome. The NNT may provide a more direct measure of clinical benefit than measures of association such as the odds ratio or the relative risk. Thus it is a useful tool for communicating results of RCTs.

Consider a randomized controlled trial with a treatment group and a placebo group, and let p_T and p_C denote the proportions who experience the disease in the treatment group and the placebo group, respectively. Then the NNT is

$$\text{NNT} = \frac{1}{p_C - p_T}.$$

Large values of $p_C - p_T$ correspond to small values of the NNT. The smaller the NNT, the greater is the clinical benefit associated with the treatment. Calculating the NNT only makes sense if there is evidence that $p_T < p_C$. If the treatment is worse than the placebo, so that $p_T > p_C$, then the NNT is negative and $-$NNT is called the **number needed to harm**, or NNH.

Activity 13.2 Calculating the NNT in the altitude sickness trial

In the trial, the proportion of trekkers who develop acute mountain sickness in the placebo group is 0.336. The corresponding proportion in the group allocated to acetazolamide is 0.119. (These proportions are reported as 34% and 12% in the Abstract.)

(a) Calculate the NNT for acetazolamide.

(b) Compare the value you obtained in part (a) with that reported in the Abstract. What do you conclude?

The **Introduction** provides the background and motivation for the study, including a review of previous work that is relevant. It is an inevitable feature of many medical papers that they contain many uncommon technical terms. You should try not to let yourself be put off by these, and keep your focus on the statistical issues. For example, the Introduction contains the phrase

> 'pharmaceutical prevention with acetazolamide is also effective despite common side effects such as parasthesias, dysgeusia, and diuresis, which may reduce compliance'.

From a statistical standpoint, it is not strictly necessary to know what parasthesias, dysgeusia, and diuresis are. All you need to know is that they are common side effects of the drug acetazolamide, and that these side effects can cause people to stop taking the drug.

Activity 13.3 Reading the Introduction

Read the Introduction, with the specific aim of identifying the following.

(a) The disease outcomes of interest (in broad terms).

(b) The treatments to be studied.

Comment

Here is a brief explanation of some of the terms and phrases used.

Acute hypobaric hypoxia: lack of oxygen due to low pressure.

Cerebral or *pulmonary oedema*: fluid build-up in the brain (cerebral) or lung (pulmonary).

Primary mediators of risk: risk factors.

Prophylactic agents: preventative treatments.

Section 13 The medical literature

13.3 The Methods section

The **Methods** section includes details of the design of the study and of the methods used to collect and analyse the data. In the first part of the Methods section, the primary outcome measure is defined and the secondary end-points are listed, along with other variables that were collected to investigate possible confounding. Details are also provided of the drugs used in the trial.

> End-point is another term for outcome.

Activity 13.4 The trial end-points

Read the first part of the Methods section, stopping just before the subsection entitled 'Randomisation and follow up'.

(a) For the purpose of this trial, how is 'acute mountain sickness' defined?

(b) Is headache on its own a primary or secondary end-point?

Activity 13.5 Randomization and follow-up

Now read the subsection entitled 'Randomisation and follow up'.

(a) Describe the groups in the trial.

(b) From the description given, was the randomization stratified?

(c) Is this an open trial, a single-blind trial or a double-blind trial?

All the disease outcomes you have met so far in this course have been discrete. The main end-points of this trial are also discrete, namely presence or absence of acute mountain sickness or headache, of different degrees of severity. However, continuous variables such as blood oxygen saturation levels were also measured. The means of these variables are compared in the treatment and placebo groups, using the two-sample t-test.

> The two-sample t-test is not described in this course.

Activity 13.6 Choice of sample size

Read the subsection entitled 'Power calculation and statistical analysis'. What further information would you need to check the sample size calculation?

Comment
The term 'asymptotic' in relation to confidence intervals means that they are based on normality assumptions that are valid in large samples. The confidence intervals described in Section 2 of this book are asymptotic confidence intervals.

Notice that, in the first sentence of the subsection entitled 'Power calculation and statistical analysis', it is stated that a sample size of 70 people per group is required 'to determine a statistically significant difference'. In fact, the authors should really have written 'to determine a practically significant difference' (and they should have stated what they took such a difference to be). Whether such a difference, if genuine, then turns out to be statistically significant depends on chance, and on the power of the trial. Such occasional minor slips in correct statistical usage are not uncommon in published papers.

13.4 The Results section

The **Results** section contains a descriptive summary of the data, together with the results of numerical calculations. Read the first paragraph of the Results section, and study the figure showing the flow chart of the trial. Also look at Table 1 of the paper. The row labels explain what data are presented. For example, 'No (%) male' in the first row means that first the number of males is given, followed by the percentage in brackets. Thus there were 337 males among the 487 participants, namely 69%.

Activity 13.7 Losses to follow-up and baseline characteristics

(a) How many participants were lost to follow-up in each group? Do these numbers vary substantially between groups?

(b) What is the average age of the 487 participants? What proportion were enrolled at an altitude of 4280 m? (*Hint*: read the footnotes to Table 1.)

(c) The authors state that the participants in the four groups were similar at baseline. In your view, do the data in Table 1 of the paper support this view?

Activity 13.8 Intention-to-treat analysis

Now read the second paragraph of the Results section (Table 2 ... pulmonary oedema).

The analysis is described as by intention to treat.

(a) Were all randomized participants included in the analysis?

(b) In what sense is this an intention-to-treat analysis?

(c) In the text, the authors refer implicitly to a per-protocol analysis. Did this yield different results from the intention-to-treat analysis?

Table 2 contains the main results of the paper, and gives odds ratios and 95% confidence intervals. The first column, labelled 'Variables', gives the outcome variable. The next column, labelled 'All participants ($n = 487$)', gives the numbers of participants with each outcome, and the percentage in brackets. For example, 115 participants had acute mountain sickness, that is, a proportion of $115/487 \simeq 0.236$, or about 24%.

The third column, which is labelled 'Placebo group ($n = 119$)', contains the numbers of participants in the placebo group with each outcome. Further to the right is a column labelled 'Ginkgo group ($n = 124$)', and to the right of this, a column containing odds ratios. These odds ratios relate to the association between each outcome and the receipt of ginkgo. For example, 40 participants in the placebo group (out of 119) had acute mountain sickness, compared to 43 (out of 124) in the ginkgo group. The odds ratio of 0.95 in the next column is for the association between *avoidance* of acute mountain sickness and treatment with ginkgo. The data are set out in Table 13.1.

Table 13.1 Ginkgo and avoidance of acute mountain sickness

Treatment group	Did not suffer from acute mountain sickness	Suffered from acute mountain sickness	Total
Ginkgo	81	43	124
Placebo	79	40	119

The odds ratio is thus

$$\widehat{OR} = \frac{81 \times 40}{43 \times 79} \simeq 0.9538.$$

This is rounded to 0.95 in Table 2 of the paper.

Section 13 The medical literature

Many of the odds ratios quoted in the paper can be reproduced directly given the counts. However, others cannot — for example, those relating to the variables Headache incidence and Non-compliant in Table 2 of the paper. In Activities 13.9 and 13.10, only odds ratios directly reproducible using the data provided are considered.

Activity 13.9 The effect of ginkgo

(a) It is stated in the text that ginkgo did not reduce the incidence of acute mountain sickness compared to placebo. On what information in Table 2 of the paper is this statement based?

(b) In Table 2 of the paper, the odds ratio for the association between treatment with ginkgo and absence of severe headache is quoted as 0.65, with 95% confidence interval (0.32, 1.29). Reproduce these results using the data in the table.

Activity 13.10 The effect of acetazolamide

Now read the rest of the Results section, from the paragraph beginning 'Acetazolamide as a control ...' to the end of the section.

(a) From Table 2 of the paper, write down the odds ratio for the association between treatment with acetazolamide and absence of acute mountain sickness, and its 95% confidence interval. What do you conclude from this?

(b) In the text the NNT for treatment of severe acute mountain sickness with acetazolamide is quoted as 7. Reproduce this result using data from Table 2 of the paper.

(c) Did the combination of acetazolamide and ginkgo provide any more protection against acute mountain sickness than acetazolamide alone? Explain your answer.

Table 3 of the paper contains comparisons between the groups at the end of the trial. One variable is labelled 'Acclimatisation nights: Midpoint (4595 m)'. As explained earlier in the paper, some participants chose to spend a night at a lodge at 4595 m, about halfway between the starting point and the end-point of the climb. These participants thus ascended more slowly, and hence had more time to acclimatize. Ascending more slowly is one way to avoid acute mountain sickness. In all, 95 participants (19%) spent a night at the halfway point. In the ginkgo group, 11% spent a night at the midpoint, compared to 22% in the acetazolamide group, 20% in the placebo group and 25% in the combined group.

See Table 3 of the paper.

Activity 13.11 Confounding by ascent rate

(a) Explain briefly how an imbalance in ascent rate between the groups might confound the association between treatment and avoidance of acute mountain sickness.

(b) What test would you use to check the authors' claim that there is no difference between the proportions spending a night at the midpoint in the four groups? Identify the data you need for the test from Table 3 in the paper. (Note: you are not asked to carry out the test.)

Note that the total for the Acetazolamide group is 118.

A chi-squared test for no association between the proportion spending a night at the midpoint and treatment group gives $\chi^2 = 7.93$, for which $p = 0.048$. Thus there is weak evidence of an imbalance in the proportions spending a night at the climb midpoint, owing mainly to the lower proportion in the ginkgo group. The authors' claim that there is no difference between the groups is thus not strictly correct. It was based on the fact that the three odds ratios relative to the placebo group were not significantly different from 1. (See Table 3 in the paper.)

Nevertheless, this imbalance between groups is unlikely to affect the results, as it is offset by the slightly higher proportion of participants in the ginkgo group who spent more than one night at baseline.

13.5 The Discussion

The final part of the paper is the **Discussion**, in which the authors summarize and discuss their results. Read this section now.

Activity 13.12 Comparison with meta-analysis

One new finding of this paper is that a 500 mg daily dose of acetazolamide is effective in preventing acute mountain sickness. The authors contrast this with the results of a meta-analysis, which found that 500 mg of acetazolamide does not work. This meta-analysis was based on three small RCTs and reported a pooled relative risk of 1.22, with 95% confidence interval (0.93, 1.59), for the prevention of acute mountain sickness with 500 mg acetazolamide daily.

(a) Use the data in Table 2 of the paper to calculate the relative risk for prevention of acute mountain sickness with 500 mg acetazolamide in the present trial, and obtain a 95% confidence interval.

(b) Do you agree with the authors' claim that their results 'clearly counter' the results of the meta-analysis?

The paper ends with a list of contributors, details of funding, competing interests and ethical approval, and the list of references. You need not read these!

In the opinion of the Course Team, this is a study of good quality, with good power, and the results therefore carry conviction, even though the report of the study is slightly marred by a few errors and obscurities. After working through this book and reading the paper, you are in a position to evaluate its findings yourself. Do you believe the results?

Summary of Section 13

In this section, you have read in detail a paper published in the medical literature. The paper made use of the following concepts and techniques covered in this book: randomized controlled trial, blinding, association, odds ratio, relative risk, confidence interval, intention-to-treat analysis, chi-squared test, Fisher's exact test, sample size calculation, confounding, meta-analysis.

14 Exercises on Book 1

Exercise 14.1   A cohort study of hospital infection

The bacterium *Staphylococcus aureus* (abbreviated to *S. aureus*) is a common cause of blood infections in patients admitted to hospital. (There is particular concern about forms of this bacterium, known as Methicillin-resistant *S. aureus* or MRSA, that are resistant to treatment with antibiotics.) These bacteria are widespread: many people carry them in the nose, without any problems. A cohort study was undertaken among patients admitted to hospital to determine whether nasal carriage of *S. aureus* on admission to hospital is a risk factor for subsequent blood infection. The data from the study are shown in Table 14.1.

Wertheim, H.F.L., Vos, M.C., Ott, A. *et al.* (2004) Risk and outcome of nosocomial *Staphylococcus aureus* bacteraemia in nasal carriers versus non-carriers. *Lancet*, **364**, 703–705.

Table 14.1   Nasal carriage of *S. aureus* and subsequent blood infection

Nasal carriage of *S. aureus* on admission to hospital	Subsequent *S. aureus* blood infection		
	Yes	No	Total
Yes	40	3 380	3 420
No	41	10 547	10 588

(a) Estimate the relative risk of blood infection in carriers compared to non-carriers.

(b) Obtain an approximate 95% confidence interval for the relative risk.

(c) Summarize and interpret your results.

Exercise 14.2   Asthma in children from three Chinese cities

In recent decades there has been a steady increase in the numbers of children suffering from asthma. Many studies have been undertaken in an attempt to understand the reasons for this increase. A cohort study was done to compare the proportions of children aged ten years who suffer from asthma in three cities in China: Hong Kong, Guangzhou, and Beijing. The data are shown in Table 14.2.

Wong, G.W.K., Ko, F.W.S., Hui, D.S.C. *et al.* (2004) Factors associated with difference in prevalence of asthma in children from three cities in China: multicentre epidemiological study. *British Medical Journal*, **329**, 486–488.

Table 14.2   Asthma in ten-year-old children in three cities in China

	Asthma		
City	Yes	No	Total
Hong Kong	179	2 931	3 110
Guangzhou	121	3 444	3 565
Beijing	159	4 068	4 227
Total	459	10 443	10 902

(a) Calculate the chi-squared test statistic for the null hypothesis of no association between city and occurrence of asthma.

(b) Test the null hypothesis of no association.

(c) What do you conclude?

Exercise 14.3 High-visibility clothing and motorcycle crashes

The inability of other road users to see motorcyclists is thought to be an important factor in motorcycle crashes. A case-control study was undertaken to investigate this hypothesis.

Wells, S., Mullin, B., Norton, R. et al. (2004) Motorcycle rider conspicuity and crash related injury: case-control study. *British Medical Journal*, **328**, 857–860.

The cases included 457 motorcycle drivers or pillion passengers who were killed or injured in a motorcycle crash. The controls included 1227 motorcycle riders selected at random from 150 roadside study sites. The exposure was wearing high-visibility (fluorescent or reflective) clothing. The data are in Table 14.3.

Table 14.3 High-visibility clothing and crash-related motorcycle injury

Wearing high-visibility clothing	Cases	Controls
Yes	49	242
No	408	985
Total	457	1227

(a) Estimate the odds ratio for the association between wearing high-visibility clothing and crash-related injury, and obtain a 95% confidence interval for the odds ratio.

(b) Interpret your results.

(c) Describe under what conditions age might confound the association between sustaining a motorcycle crash injury and wearing high-visibility clothing. The authors of the study reported that, after adjusting for age, the odds ratio was 0.50, with 95% confidence interval (0.36, 0.70). Did confounding by age arise in this study? Justify your answer.

Exercise 14.4 Physical activity and heart attacks

A 1–1 matched case-control study was undertaken to investigate the hypothesis that physical activity protects against myocardial infarction (heart attacks). Cases were individuals who suffered a heart attack. For each case, a control was selected of the same sex, living in the same area, and of similar age (within five years). Exposure was defined as taking sufficient physical exercise to expend 2500 kilocalories per day on average. The data, on 340 case-control pairs, are in Table 14.4.

Hennekens, C.H. and Buring, J.E. (1987) *Epidemiology in Medicine*. Lippincott-Raven, Philadelphia.

Table 14.4 Distribution of case-control pairs

		Controls	
		Exposed	Not exposed
Cases	Exposed	115	59
	Not exposed	99	67

(a) Estimate the Mantel–Haenszel odds ratio for the association between taking physical exercise and suffering a heart attack, and obtain a 95% confidence interval for the odds ratio.

(b) Carry out McNemar's test for no association.

(c) Interpret your results.

Exercise 14.5 A case-control study of stress and heart disease

A case-control study was carried out to investigate the association between stress and heart disease. The cases were people who had suffered a myocardial infarction, and the controls were people of similar ages without heart disease. Cases and controls were asked about history of stress in various settings.

The data for stress at work for males are in Table 14.5. Stress levels have been grouped into four categories: never stressed, stressed some of the time, stressed often, always stressed.

Rosengren, A., Hawken, S., Ôunpuu, S. et al. (2004) Association of psychosocial risk factors with risk of acute myocardial infarction in 11 119 cases and 13 648 controls from 52 countries (the INTERHEART study): case-control study. *Lancet*, **364**, 953–962.

Table 14.5 Myocardial infarction and stress at work

Stress at work	Cases	Controls
Always	499	316
Often	1125	1117
Sometimes	2265	3315
Never	993	1504
Total	4882	6252

(a) Calculate the dose-specific odds ratios relative to the Never category. Discuss any patterns you observe.

(b) The test statistic for the chi-squared test for no linear trend is 114.11. Calculate the p value for the test, and interpret your findings.

Exercise 14.6 Treatment of tetanus: randomized controlled trial

Tetanus is a serious, often fatal disease of the nervous system caused by infection of a wound with the bacterium *Clostridium tetani*. A randomized controlled trial was undertaken in Brazil of two treatments for tetanus by injection of a drug (antitetanus immunoglobulin). In the treatment group, the drug was injected both into the spinal fluid and into muscle. In the control group, the drug was injected only into muscle.

Miranda-Filho, D.B., Ximenes, R.A.A., Barone, A.A. et al (2004) Randomised controlled trial of tetanus treatment with antitetanus immunoglobulin by the intrathecal or intramuscular route. *British Medical Journal*, **328**, 615–617.

(a) Randomization was by blocks of 20. Explain what this means.

(b) For two patients allocated to the treatment group, it was not possible to administer the injection into spinal fluid. However, in the analysis, these two patients were counted in the treatment group, even though they received the same care as those allocated to the control group. What is this method of analysis called? Briefly explain its rationale.

(c) The power of the trial turned out to be too low to evaluate reliably the effect of the treatment on mortality. Calculate the sample size that would be required in each group to detect a reduction of 10% in the proportion who die, with 80% power and 5% significance level, if the death rate in the control group is 15%.

Exercise 14.7 MMR vaccine and autism: meta-analysis

In Example 7.3, a case-control study to investigate the hypothesis that measles, mumps and rubella (MMR) vaccine causes autism was described. In the same paper, the authors undertook a meta-analysis of four of the epidemiological studies that have been published on this issue. The forest plot is shown in Figure 14.1.

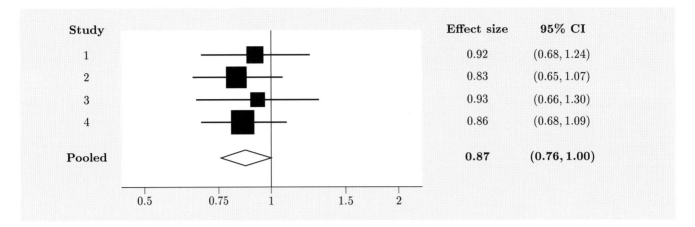

Figure 14.1 A forest plot for four studies on MMR vaccine and autism

The four studies included two cohort studies (studies 1 and 2) and two case-control studies (studies 3 and 4). The two cohort studies reported relative risks, and the two case-control studies reported odds ratios. These are represented together on the plot, under the common label 'Effect size'. Values greater than 1 indicate that autism is positively associated with receipt of the MMR vaccine.

(a) Autism is a rare condition. Explain why it is appropriate to combine odds ratios and relative risks as the authors have done.

(b) Which study contributed most to the overall estimate? Which study contributed least?

(c) What do you conclude from this meta-analysis?

Summary of Book 1

Part I

Cohort studies and case-control studies are contrasting study designs for evaluating the association between an exposure and a disease. In a cohort study, individuals are classified according to their exposure and followed forward in time to evaluate disease outcomes. In contrast, in a case-control study, individuals are selected according to their disease status; their past exposures are then ascertained. In both types of study, the strength of association may be quantified by the odds ratio. In a cohort study, the relative risk may also be estimated. Uncertainty in the odds ratio or relative risk, and hence in the strength of association, may be summarized by a confidence interval. In addition, the chi-squared test or Fisher's exact test may be used to test the null hypothesis of no association.

Part II

Selection bias, information bias and confounding may result in incorrect inferences about an association. Little can be done to correct selection bias and information bias if they are present. However, it may be possible to remove the effect of confounding, which arises if a third variable is associated with both the exposure and the disease. This is achieved by stratifying the data according to the levels of this third variable. The Mantel–Haenszel method provides one approach to estimating the odds ratio for stratified data. Matching is a form of stratification often used in case-control studies; and the analysis of 1–1 matched studies can be undertaken using McNemar's test. The Mantel–Haenszel method and McNemar's test require the assumption that the odds ratio is the same across strata. This may be tested using Tarone's test for homogeneity. When bias is deemed to be an unlikely explanation for an observed association, Bradford Hill's criteria provide a framework for evaluating the evidence that the association is causal. Such evidence is strengthened if a dose-response relationship can be demonstrated. The chi-squared test for no linear trend can be used to investigate the presence of a dose-response relationship.

Part III

A randomized controlled trial (RCT) is a special type of cohort study, in which randomization and concealment are used to eliminate bias. Different schemes are available, such as stratified randomization and double blinding. Flow charts are used to document the progress of the trial, in particular the numbers of participants lost to follow-up and protocol violations. In order to avoid selection bias, RCTs should usually be analysed using the intention-to-treat principle. Designing an RCT involves formulating the trial hypotheses and calculating the sample size required to achieve a pre-specified significance level and power. The results of several RCTs or epidemiological studies may be evaluated together in a systematic review. Meta-analysis may also be used to pool the results of several studies, and their results may be contrasted using a forest plot. The methods described in this book are commonly used in the modern medical literature.

Learning outcomes

You have been working to develop the following skills.

Part I

- Describe cohort studies and case-control studies for investigating the association between an exposure E and a disease D.
- Describe the use of relative risks and odds ratios as measures of association.
- Estimate relative risks and odds ratios using data from a cohort study.
- Estimate odds ratios using data from a case-control study.
- Calculate and interpret confidence intervals for the relative risk and odds ratio.
- Test for no association using the chi-squared test.
- Interpret the results of a chi-squared test for no association and Fisher's exact test.
- Use SPSS to tabulate and analyse data from cohort studies and case-control studies.

Part II

- Describe the types of bias known as selection bias, information bias and confounding, and the procedure of matching in a case-control study.
- Calculate and interpret the Mantel–Haenszel estimate of the odds ratio for stratified tables.
- Calculate and interpret the Mantel–Haenszel estimate of the odds ratio and a confidence interval for the odds ratio for a 1–1 matched case-control study.
- Conduct McNemar's test for no association in a 1–1 matched case-control study and interpret the results.
- Identify interactions and interpret Tarone's test for homogeneity for stratified tables.
- Investigate dose-response relationships and interpret the results of a chi-squared test for no linear trend.
- Use SPSS to enter and analyse stratified tabular data, and to analyse a 1–1 matched case-control study.

Part III

- Describe the role of randomization in randomized controlled trials, and the method of stratified randomization by blocks.
- Describe the role of concealment in randomized controlled trials, and identify double-blind, open, and single-blind trials.
- Draw and interpret the flow chart for a randomized controlled trial.
- Describe the intention-to-treat and per-protocol analysis methods, and analyse a trial by intention to treat.
- Describe the four phases of drug evaluation in humans and the role of data monitoring committees.
- Formulate the trial hypotheses and specify the significance level, power, and design values for a trial.
- Calculate the sample size required for a trial, and the power available for a given sample size.
- Describe the purpose of systematic reviews and meta-analyses.
- Interpret the forest plot for a meta-analysis.
- Read critically and interpret suitable articles in the medical literature.

Solutions to Activities

Solution 1.1

(a) The exposure E is compulsory redundancy. The disease D is serious self-inflicted injury leading to hospitalization or death.

(b) The exposed group comprises the Whakatu workers. The control group comprises the Tomoana workers.

(c) The data from the study can be arranged as shown in Table S.1.

Table S.1 Serious self-inflicted injury (SSII) and compulsory redundancy in meat-processing workers in New Zealand, 1986–94

Exposure category	SSII	No SSII	Total
Made compulsorily redundant (Whakatu workers)	14	1931	1945
Not made compulsorily redundant (Tomoana workers)	4	1763	1767

Solution 1.2

(a) The estimated probabilities are
$$\widehat{P}(D|E) = \frac{14}{33} \simeq 0.42$$
and
$$\widehat{P}(D|\text{not } E) = \frac{13}{52} = 0.25.$$

(b) The estimated probability that a child sustains at least moderately severe injury is higher for children not wearing a seat belt than for children wearing a seat belt. However, it cannot be concluded that $P(D|E)$ is greater than $P(D|\text{not } E)$ in the population, since the observed difference might be due to random variation.

Solution 1.3

(a) The estimated relative risk is
$$\widehat{RR} = \frac{a/n_1}{c/n_2} = \frac{39/52}{19/33} \simeq 1.30.$$

(b) This estimate is less than the value 1.70 obtained in Example 1.3.

(c) Wearing a seat belt is associated with a 30% increase in the chance of avoiding moderate or worse injury. The strength of association between seat belt use and moderate or worse injury has not changed, but our measure of it has.

Solution 1.4

Using Formula (1.2),
$$\widehat{OR} = \frac{a \times d}{b \times c} = \frac{39 \times 14}{13 \times 19} \simeq 2.21.$$

This value is the same as that obtained for $\widehat{OR}$ in Example 1.4.

Solution 1.5

(a) The estimated relative risk is
$$\widehat{RR} = \frac{a/n_1}{c/n_2} = \frac{14/1945}{4/1767} \simeq 3.18.$$
The estimated odds ratio is
$$\widehat{OR} = \frac{a \times d}{b \times c} = \frac{14 \times 1763}{1931 \times 4} \simeq 3.20.$$
The estimated relative risk and the estimated odds ratio are both greater than 1. Thus both measures indicate that compulsory redundancy may be positively associated with SSII.

(b) The estimated relative risk and the estimated odds ratio are very similar. Serious self-inflicted injury is uncommon, so in this case the relative risk and the odds ratio nearly coincide.

(c) For a non-statistical audience the terms 'relative risk' and 'odds ratio' should be avoided. So an appropriate description of the results, which implicitly uses relative risks, is as follows.

'Workers who were made compulsorily redundant were three times more likely to suffer serious self-inflicted injury, compared to workers who were not made redundant.'

Solution 2.1

(a) The two groups are drivers with air bags, and passengers without air bags. Thus the effects of seat position and air bag use cannot be separated. The data suggest that the risk of death is lower for drivers with air bags compared to passengers without air bags (0.53 compared to 0.65). However, it is not possible, from these data alone, to infer whether this is due to an association with seat position, air bags, or both.

(b) The random variables X and Y denoting numbers of fatalities in the two groups are not independent, since the data are obtained from driver-passenger pairs. For example, in very severe crashes there is a higher probability that both driver and passenger will be killed than in less severe crashes.

Solution 2.2

(a) The estimated relative risk is given by (2.2):
$$\widehat{RR} = \frac{a/n_1}{c/n_2} = \frac{156/9577}{1531/16\,328} \simeq 0.1737.$$

(b) The estimated standard error $\hat{\sigma}$ of RR is given by (2.4):
$$\hat{\sigma} = \sqrt{\frac{1}{a} - \frac{1}{n_1} + \frac{1}{c} - \frac{1}{n_2}}$$
$$= \sqrt{\frac{1}{156} - \frac{1}{9577} + \frac{1}{1531} - \frac{1}{16\,328}}$$
$$\simeq 0.08305.$$

For a 99% confidence interval, the 0.995-quantile of the standard normal distribution is required; this is $z = 2.576$. The confidence limits are given by (2.3):

$$RR^- = \widehat{RR} \times \exp(-z \times \widehat{\sigma})$$
$$\simeq 0.1737 \times \exp(-2.576 \times 0.08305)$$
$$\simeq 0.14,$$
$$RR^+ = \widehat{RR} \times \exp(z \times \widehat{\sigma})$$
$$\simeq 0.1737 \times \exp(2.576 \times 0.08305)$$
$$\simeq 0.22.$$

So a 99% confidence interval for RR is $(0.14, 0.22)$.

(c) The relative risk is 0.17, with 99% confidence interval $(0.14, 0.22)$. The estimate of RR and its confidence interval are located well below 1, indicating a negative association between measles vaccination and measles infection.

Solution 2.3

(a) The estimated odds ratio is
$$\widehat{OR} = \frac{a \times d}{b \times c} = \frac{20 \times 534}{126 \times 51} \simeq 1.6620.$$

(b) To calculate a confidence interval, the estimated standard error $\widehat{\sigma}$ is required:
$$\widehat{\sigma} = \sqrt{\frac{1}{a} + \frac{1}{b} + \frac{1}{c} + \frac{1}{d}}$$
$$= \sqrt{\frac{1}{20} + \frac{1}{126} + \frac{1}{51} + \frac{1}{534}}$$
$$\simeq 0.2818.$$

For a 95% confidence interval, the 0.975-quantile of the standard normal distribution is required, namely $z = 1.96$. The 95% confidence limits are
$$OR^- = \widehat{OR} \times \exp(-z \times \widehat{\sigma})$$
$$\simeq 1.6620 \times \exp(-1.96 \times 0.2818)$$
$$\simeq 0.96,$$
$$OR^+ = \widehat{OR} \times \exp(z \times \widehat{\sigma})$$
$$\simeq 1.6620 \times \exp(1.96 \times 0.2818)$$
$$\simeq 2.89.$$

So a 95% confidence interval for OR is $(0.96, 2.89)$.

(c) The odds ratio is 1.66, with 95% confidence interval $(0.96, 2.89)$. The confidence interval includes 1. This means that we cannot conclude that there is an association. Note, however, that this study does not rule out such an association: the data are also consistent with values of OR that are greater than 1.

Solution 3.1

(a) The exposure in this study is attendance at political meetings. A table similar to Table S.2 is required.

Table S.2 Homicide and political activity in Karachi

Exposure category	Cases	Controls
Attended political meetings	11	2
Did not attend political meetings	24	83
Total	35	85

(b) The proportion exposed in cases is $11/35 \simeq 0.31$ and in controls is $2/85 \simeq 0.02$. Thus the proportion exposed (that is, who attended political meetings) is considerably higher in cases (homicide victims) than in controls. This suggests that attending political meetings might be positively associated with death by homicide.

Solution 3.2

(a) The estimated odds ratio is
$$\widehat{OR} = \frac{a \times d}{b \times c} = \frac{62 \times 55}{76 \times 5} \simeq 8.9737.$$

(b) For a 95% confidence interval, the 0.975-quantile of the standard normal distribution is required, namely $z = 1.96$. The estimated standard error $\widehat{\sigma}$ is given by
$$\widehat{\sigma} = \sqrt{\frac{1}{a} + \frac{1}{b} + \frac{1}{c} + \frac{1}{d}}$$
$$= \sqrt{\frac{1}{62} + \frac{1}{76} + \frac{1}{5} + \frac{1}{55}}$$
$$\simeq 0.4975.$$

So the confidence limits are
$$OR^- = \widehat{OR} \times \exp(-z \times \widehat{\sigma})$$
$$\simeq 8.9737 \times \exp(-1.96 \times 0.4975)$$
$$\simeq 3.38,$$
$$OR^+ = \widehat{OR} \times \exp(z \times \widehat{\sigma})$$
$$\simeq 8.9737 \times \exp(1.96 \times 0.4975)$$
$$\simeq 23.79.$$

So a 95% confidence interval for the odds ratio is $(3.38, 23.79)$.

(c) The estimated odds ratio is 8.97. This means that the odds of SIDS for infants placed to sleep on their front is 8.97 times the odds of SIDS for infants laid down to sleep in other positions. The 95% confidence interval is $(3.38, 23.79)$. This lies well above 1. In conclusion, the data indicate that there exists a positive association between death from SIDS and putting the baby down to sleep on its front.

Solutions to Activities

Solution 3.3

(a) A good choice of reference category is women with no history of genital infections, though the choice is to some extent arbitrary.

(b) Relative to this category, the estimated odds ratio for PID is
$$\widehat{OR} = \frac{212 \times 1154}{112 \times 411} \simeq 5.3147.$$
In this case,
$$\hat{\sigma} = \sqrt{\frac{1}{212} + \frac{1}{112} + \frac{1}{411} + \frac{1}{1154}} \simeq 0.1302.$$
So the approximate 95% confidence limits are
$$OR^- \simeq 5.3147 \times \exp(-1.96 \times 0.1302) \simeq 4.12,$$
$$OR^+ \simeq 5.3147 \times \exp(1.96 \times 0.1302) \simeq 6.86.$$
Hence the 95% confidence interval is $(4.12, 6.86)$.

The odds ratio for Non-PID infections relative to None is
$$\widehat{OR} = \frac{157 \times 1154}{407 \times 411} \simeq 1.0831.$$
In this case,
$$\hat{\sigma} = \sqrt{\frac{1}{157} + \frac{1}{407} + \frac{1}{411} + \frac{1}{1154}} \simeq 0.1101.$$
So the 95% confidence limits for this odds ratio are
$$OR^- \simeq 1.0831 \times \exp(-1.96 \times 0.1101) \simeq 0.87,$$
$$OR^+ \simeq 1.0831 \times \exp(1.96 \times 0.1101) \simeq 1.34.$$
Thus the 95% confidence interval is $(0.87, 1.34)$.

(c) There is a positive association between a pregnancy being ectopic and PID: the odds ratio is 5.31, and the 95% confidence interval $(4.12, 6.86)$ is located well above 1. However, for infections other than PID, there is little evidence of any association with ectopic pregnancy: the odds ratio is only 1.08, and the 95% confidence interval $(0.87, 1.34)$ contains 1.

Solution 4.1

(a) The frequencies expected under the null hypothesis of no association are given in Table S.3.

Table S.3 Expected frequencies for SIDS data

Position baby last placed down to sleep	Cases	Controls	Total
On its front	10.55	43.45	54
On its side	61.95	255.05	317
On its back	115.50	475.50	591
Total	188	774	962

For example, the expected frequency of cases among babies last placed on their front is given by
$$\frac{\text{row total} \times \text{column total}}{\text{overall total}} = \frac{54 \times 188}{962} \simeq 10.55.$$

(b) The numbers of cases observed are greater than expected for babies placed on their front or side. This suggests that there might be a positive association between SIDS and placing a baby down on its front or side. (However, this does not on its own prove the existence of such an association.)

Solution 4.2

The observed frequencies O_i are in Table 4.4 and the expected frequencies E_i are in Table S.3. The value of the chi-squared test statistic is given by
$$\chi^2 = \sum_i \frac{(O_i - E_i)^2}{E_i}$$
$$\simeq \frac{(30 - 10.55)^2}{10.55} + \frac{(76 - 61.95)^2}{61.95} + \frac{(82 - 115.50)^2}{115.50}$$
$$+ \frac{(24 - 43.45)^2}{43.45} + \frac{(241 - 255.05)^2}{255.05} + \frac{(509 - 475.50)^2}{475.50}$$
$$\simeq 35.8581 + 3.1865 + 9.7165 + 8.7066 + 0.7740 + 2.3601$$
$$\simeq 60.60.$$

Solution 4.3

(a) The 0.95-quantile of $\chi^2(5)$ is 11.07.

(b) The value required is the 0.975-quantile of $\chi^2(8)$, namely 17.53.

(c) Looking along the row of the table corresponding to $\nu = 3$, the value 10.25 lies between the 0.975-quantile (9.35) and the 0.99-quantile (11.34). Thus
$$0.01 < P(W > 10.25) < 0.025.$$
Hence, from the table, the best lower bound for the probability is 0.01 and the best upper bound is 0.025. (The value of the probability is, in fact, approximately 0.01656.)

Solution 4.4

(a) To calculate the p value, the null distribution of the test statistic is required. Since Table 4.5 is a 3×2 table, the null distribution is approximately chi-squared with degrees of freedom $\nu = (3-1) \times (2-1) = 2$. Since the expected frequencies are all greater than 5, the approximation is adequate.

The 0.80-quantile of $\chi^2(2)$ is 3.22, which is greater than 3.11, the observed value of χ^2. Thus the p value is greater than 0.2.

(b) Since $p > 0.2$, there is little evidence of association between childhood asthma and gestational age.

Solution 4.5

The expected frequencies under the null hypothesis of no association are in Table S.4.

Table S.4 Ectopic pregnancy and history of genital infections — expected frequencies

History	Cases	Controls	Total
PID	103.02	220.98	324
Non-PID	179.34	384.66	564
None	497.64	1067.36	1565
Total	780	1673	2453

The value of the chi-squared test statistic is
$$\chi^2 = \frac{(212-103.02)^2}{103.02} + \frac{(157-179.34)^2}{179.34}$$
$$+ \frac{(411-497.64)^2}{497.64} + \frac{(112-220.98)^2}{220.98}$$
$$+ \frac{(407-384.66)^2}{384.66} + \frac{(1154-1067.36)^2}{1067.36}$$
$$\simeq 195.23.$$

The null distribution of the test statistic is approximately chi-squared with degrees of freedom $\nu = (3-1) \times (2-1) = 2$. Since all expected frequencies are at least 5, the approximation is adequate.

The 0.999-quantile of the $\chi^2(2)$ distribution is 13.82. The observed value of the test statistic is greater than this, so the p value is less than 0.001. There is strong evidence of an association between prior genital infection and ectopic pregnancy.

Solution 4.6

(a) The expected frequencies are shown in brackets in Table S.5.

Table S.5 Observed and expected frequencies for lung cancer data

Exposure category	Cases	Controls	Total
Smoked	647 (634.50)	622 (634.50)	1269
Never smoked	2 (14.50)	27 (14.50)	29
Total	649	649	1298

The observed value of the chi-squared test statistic is
$$\chi^2 = \sum_i \frac{(O_i - E_i)^2}{E_i}$$
$$= \frac{(647-634.50)^2}{634.50} + \frac{(622-634.50)^2}{634.50}$$
$$+ \frac{(2-14.50)^2}{14.50} + \frac{(27-14.50)^2}{14.50}$$
$$\simeq 0.2463 + 0.2463 + 10.7759 + 10.7759$$
$$\simeq 22.04.$$

The null distribution of the test statistic is approximately chi-squared with degrees of freedom $\nu = (2-1) \times (2-1) = 1$. Since the expected frequencies are all at least 5, the approximation is adequate. (Note that one *observed* value is less than 5, but this does not matter.)

Using tables, the 0.999-quantile of $\chi^2(1)$ is 10.83. Since the value of the test statistic is greater than this, the p value is less than 0.001. There is strong evidence of an association between smoking and lung cancer.

(b) The estimated odds ratio is
$$\widehat{OR} = \frac{647 \times 27}{622 \times 2} \simeq 14.0426 \simeq 14.04.$$
To calculate an approximate 95% confidence interval, the estimated standard error $\hat\sigma$ is required:
$$\hat\sigma = \sqrt{\frac{1}{647} + \frac{1}{622} + \frac{1}{2} + \frac{1}{27}} \simeq 0.7350.$$
The 95% confidence limits are
$$OR^- \simeq 14.0426 \times \exp(-1.96 \times 0.7350) \simeq 3.33,$$
$$OR^+ \simeq 14.0426 \times \exp(1.96 \times 0.7350) \simeq 59.31.$$
So the 95% confidence interval for the odds ratio is $(3.33, 59.31)$.

(c) There is strong evidence of an association between smoking and lung cancer ($p < 0.001$). The estimated odds ratio is 14.04, with 95% confidence interval $(3.33, 59.31)$. This indicates a strong positive association. In conclusion, this study provides *strong evidence* of a *strong positive association* between smoking and lung cancer.

Solution 6.1

(a) The main difference between the selection procedures arose when several potential controls were available; in such cases, the first on the ward list was chosen.

(b) If non-smokers were placed first on the ward lists, then if several controls were available, a non-smoker would be more likely to be chosen. However, ordering the ward lists in this way seems very unlikely, especially since smoking was not widely recognized as a health risk at the time. Could smokers with lung cancer unwittingly be over-represented in the study hospitals — for example, owing to non-smokers with lung cancer (but not smokers with other diseases) preferring to get treated at other hospitals? This also appears rather far-fetched, though not totally impossible.

It seems reasonable to conclude that it is unlikely that selection bias could account for the high odds ratio.

(c) Suppose that the controls could be divided into two groups: controls whose disease is caused by smoking, and 'true' controls whose disease is not caused by smoking. Then it is likely that the proportion of smokers among the 'true' controls would be lower than among the other controls. If the 'true' controls were used to calculate the odds ratio for lung cancer, then it is likely that the value obtained would be higher than the value observed. Thus the odds ratio may be underestimated in this study.

Solutions to Activities

Solution 6.2

(a) The same questionnaire was used for cases and controls, and the same procedures were followed by interviewing a relative of the study subject. So this aspect of the data collection is unlikely to be biased.

(b) The context of the study must be taken into account. The families of cases might feel less threatened by reporting political activity than families of controls (who are still alive), especially if participation in political meetings is a risk factor for death by homicide.

(c) Thus information bias could be a factor, and it is perhaps likely to bias the odds ratio upwards.

Solution 6.3

(a) The stratum-specific odds ratios are as follows:
$$\widehat{OR}_{C=0} = \frac{70 \times 6}{30 \times 14} = 1, \quad \widehat{OR}_{C=1} = \frac{14 \times 78}{26 \times 42} = 1.$$
Since both odds ratios are equal to 1, there is no association between E and D.

(b) The aggregated data are in Table S.6.

Table S.6 Data aggregated over the levels of C

	Disease	No disease
Exposed	84	56
Not exposed	56	84

(c) The odds ratio from the aggregated data is given by
$$\widehat{OR} = \frac{84 \times 84}{56 \times 56} = 2.25.$$

(d) The conclusion is that the apparent association between E and D with an odds ratio of 2.25 from the aggregated data is the result of confounding by variable C: if the data are stratified by the levels of C, then the association disappears. In this instance, confounding has produced a spurious association where there is none.

Solution 6.4

(a) The age-specific odds ratios are as follows:
$$\widehat{OR}_{\leq 40} = \frac{0 \times 129}{15 \times 1} = 0.00,$$
$$\widehat{OR}_{>40} = \frac{218 \times 124}{311 \times 104} \simeq 0.84.$$

(b) The data have been aggregated to give Table S.7.

Table S.7 Diabetes type and mortality: aggregated data

Diabetes type	Died	Alive	Total
Non-insulin dependent	218	326	544
Insulin dependent	105	253	358

(c) The odds ratio for the aggregated data is as follows:
$$\widehat{OR} = \frac{218 \times 253}{326 \times 105} \simeq 1.61.$$

(d) The aggregated data give an odds ratio of 1.61, suggesting that non-insulin dependent diabetes is associated with higher mortality than insulin dependent diabetes. However, the age-stratified analysis gives odds ratios of 0.00 and 0.84, both of which are less than 1. Thus non-insulin dependent diabetes is associated with lower mortality than insulin dependent diabetes. Age is a confounder: it is associated with the outcome (most deaths among diabetes patients occur at ages > 40 years) and with the exposure (non-insulin dependent diabetes is more frequent in diabetes patients aged > 40 years). The confounding is such that it reverses the direction of association.

Solution 7.1

(a) The Mantel–Haenszel odds ratio is given by Formula (7.1):
$$\widehat{OR}_{MH} = \frac{(234 \times 6)/357 + (55 \times 71)/343}{(36 \times 81)/357 + (25 \times 192)/343}$$
$$\simeq \frac{15.3176}{22.1622}$$
$$\simeq 0.69.$$

(b) As expected, the Mantel–Haenszel odds ratio lies between the stratum-specific odds ratios of 0.48 and 0.81, and is quite different from the crude odds ratio of 1.34.

Solution 7.2

(a) The Mantel–Haenszel odds ratio is given by Formula (7.1):
$$\widehat{OR}_{MH} = \frac{\sum a_i d_i / N_i}{\sum b_i c_i / N_i}$$
$$= \frac{(5 \times 23)/61 + \cdots + (5 \times 25)/75}{(25 \times 8)/61 + \cdots + (16 \times 29)/75}$$
$$\simeq \frac{4.8019}{19.3502}$$
$$\simeq 0.25.$$

(b) The aggregated data are in Table S.8.

Table S.8 Fluoridation and caries: aggregated data

Water type	With caries	Without caries	Total
Fluoridated	15	71	86
Not fluoridated	78	95	173

The crude odds ratio is given by
$$\widehat{OR} = \frac{15 \times 95}{71 \times 78} \simeq 0.26.$$

(c) The Mantel–Haenszel odds ratio for the association between water fluoridation and dental caries is 0.25, suggesting that water fluoridation protects against caries. The crude odds ratio is 0.26. This is very close to the adjusted value, suggesting that the association is not confounded by age.

Solution 7.3

(a) A table similar to Table S.9 is required.

Table S.9 Diabetes and heart attacks among Navajo Indians

		Controls	
		Exposed	Not exposed
Cases	Exposed	9	37
	Not exposed	16	82

(b) The Mantel–Haenszel odds ratio for the association between diabetes and heart attacks, adjusted for the matching factors age, sex and hospital, is obtained using Formula (7.3):
$$\widehat{OR}_{MH} = \frac{f}{g} = \frac{37}{16} = 2.3125 \simeq 2.31.$$

Solution 7.4

(a) The estimated standard error is given by (7.4). For the data in Table 7.8, $f = 37$ and $g = 16$, so
$$\hat{\sigma} = \sqrt{\frac{1}{f} + \frac{1}{g}} = \sqrt{\frac{1}{37} + \frac{1}{16}} \simeq 0.2992.$$

For a 95% confidence interval, $z = 1.96$. Thus the 95% confidence limits for the odds ratio are
$$OR^- = \widehat{OR}_{MH} \times \exp\left(-z\sqrt{\frac{1}{f} + \frac{1}{g}}\right)$$
$$\simeq 2.3125 \times \exp(-1.96 \times 0.2992)$$
$$\simeq 1.29,$$
$$OR^+ = \widehat{OR}_{MH} \times \exp\left(z\sqrt{\frac{1}{f} + \frac{1}{g}}\right)$$
$$\simeq 2.3125 \times \exp(1.96 \times 0.2992)$$
$$\simeq 4.16.$$

So the 95% confidence interval is $(1.29, 4.16)$.

(b) The observed value of McNemar's test statistic is, using (7.5),
$$\chi^2 = \frac{(|37 - 16| - 1)^2}{37 + 16} \simeq 7.55.$$

This lies between the 0.99-quantile (6.63) and the 0.995-quantile (7.88) of $\chi^2(1)$. Thus $0.005 < p < 0.01$. This constitutes strong evidence that diabetes is positively associated with heart attacks among Navajo Indians.

Solution 7.5

Since there are two strata, the null distribution of Tarone's test statistic is chi-squared with degrees of freedom $\nu = 2 - 1 = 1$. The 0.999-quantile of $\chi^2(1)$ is 10.83. Since the observed value of the test statistic is 13.77, which is greater than 10.83, $p < 0.001$. So there is strong evidence that the odds ratio differs between strata.

Solution 8.1

(a) The estimated odds ratio for the Moderate category relative to the Minimal category is given by
$$\widehat{OR} = \frac{180 \times 2749}{2783 \times 95} \simeq 1.87.$$

The other odds ratios are estimated in the same way. The values are given in Table S.10.

Table S.10 Odds ratios for stress level and PTSD

Stress level	PTSD	No PTSD	Odds ratio
Extreme	174	595	8.46
Very severe	260	1155	6.51
Severe	362	2688	3.90
High	97	1286	2.18
Moderate	180	2783	1.87
Minimal	95	2749	1.00

(b) The estimated odds ratio increases with increasing stress level. This suggests that there is an increasing dose-response relationship between stress level and PTSD. This is consistent with a causal link in the context of the 1991 Gulf War.

Solution 8.2

(a) The estimated odds of disease (PTSD/No PTSD) and their logarithms are shown in Table S.11.

Table S.11 Odds of PTSD and their logarithms by stress level

i (Stress level)	PTSD	No PTSD	$\widehat{OD}_i$	$\log(\widehat{OD}_i)$
6 (Extreme)	174	595	0.2924	−1.23
5 (Very severe)	260	1155	0.2251	−1.49
4 (Severe)	362	2688	0.1347	−2.00
3 (High)	97	1286	0.0754	−2.58
2 (Moderate)	180	2783	0.0647	−2.74
1 (Minimal)	95	2749	0.0346	−3.36

The plot of the logarithm of the estimated odds of PTSD against stress level is shown in Figure S.1.

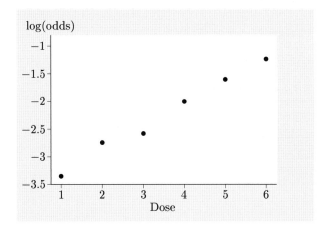

Figure S.1 A scatterplot of the logarithm of the estimated odds of PTSD against stress level

The plot suggests an increasing, linear dose-response relationship between stress level and PTSD.

Solutions to Activities

(b) The observed value of the test statistic is 455.15. The 0.999-quantile of $\chi^2(1)$ is 10.83, so $p < 0.001$. There is strong evidence for a linear dose-response relationship between stress level and post-traumatic stress disorder.

Solution 10.1

(a) The maximum difference is 3. For each block of six, three patients are allocated to each treatment. Thus any difference comes from an incomplete final block, and such a difference cannot exceed 3.

(b) Age is associated with the long-term outcome of Guillain–Barré syndrome. If age was also associated with treatment, there would be potential for confounding by age. Stratifying the randomization by age ensures that the two treatment groups are balanced with respect to age group.

Solution 10.2

The block size k should be divisible by 4. Within each block $k/4$ patient numbers should be allocated to each group in random order. For example, the block size 16 may be chosen. Then within each block, four patient numbers would be allocated to each group in random order.

Solution 10.3

(a) To conceal the allocation from the patient a placebo would have to be found with the same appearance (and taste) as the dissolved charcoal. Since such a placebo was not available the allocation could not be concealed from the patient. Since the charcoal was dissolved in water, water was also given to patients randomized to the control group to maximize comparability between the groups.

(b) The staff administering the treatment would be aware of what they were administering. Thus different trial staff would be required to carry out the patient assessments.

Solution 10.4

(a) The intention-to-treat analysis should be based on the treatments allocated, not on the treatments received. The data for the intention-to-treat analysis are in Table S.12.

Table S.12 Data for intention-to-treat analysis of yellow oleander poisoning trial

Trial group	Died	Alive	Total randomized
Treatment	5	196	201
Control	16	184	200

(b) The estimated odds ratio is
$$\widehat{OR} = \frac{5 \times 184}{196 \times 16} \simeq 0.2934 \simeq 0.29.$$
The estimated standard error $\widehat{\sigma}$ is given by
$$\widehat{\sigma} = \sqrt{\frac{1}{5} + \frac{1}{196} + \frac{1}{16} + \frac{1}{184}} \simeq 0.5225.$$
The 95% confidence limits are
$$OR^- \simeq 0.2934 \times \exp(-1.96 \times 0.5225) \simeq 0.11,$$
$$OR^+ \simeq 0.2934 \times \exp(1.96 \times 0.5225) \simeq 0.82.$$
So the odds ratio is 0.29, with 95% confidence interval $(0.11, 0.82)$.

(c) The odds ratio and confidence interval indicate that supplementary charcoal therapy is effective in reducing deaths from yellow oleander poisoning.

Solution 10.5

The study is a Phase I study: the sample size is very small (only 3); it is an uncontrolled study; and it is undertaken in adult volunteers.

Solution 10.6

(a) The analysis includes all the 345 women for whom data are available and, for an intention-to-treat analysis, is based (as far as possible) on treatment allocated, not treatment received. The data for the analysis are in Table S.13.

Table S.13 Data for interim intention-to-treat analysis of HRT trial

Trial group	New breast cancer	No new breast cancer	Total with data available
HRT	26	148	174
No HRT	8	163	171

(b) The estimated odds ratio is
$$\widehat{OR} = \frac{26 \times 163}{148 \times 8} \simeq 3.5794.$$
The estimated standard error is given by
$$\widehat{\sigma} = \sqrt{\frac{1}{26} + \frac{1}{148} + \frac{1}{8} + \frac{1}{163}} \simeq 0.4199.$$
Hence the 95% confidence limits are
$$OR^- \simeq 3.5794 \times \exp(-1.96 \times 0.4199) \simeq 1.57,$$
$$OR^+ \simeq 3.5794 \times \exp(1.96 \times 0.4199) \simeq 8.15.$$
So the 95% confidence interval is $(1.57, 8.15)$.

(c) The odds ratio is 3.58, with 95% confidence interval $(1.57, 8.15)$. The confidence interval is located well above 1. This indicates a rather strong positive association between HRT and the development of new breast cancer. It would be unethical to continue the trial. Therefore the committee should recommend that the trial be stopped.

Solution 11.1

(a) The treatment group will include patients randomized to standard therapy plus Infliximab. The control group will include patients randomized to standard therapy plus a placebo.

(b) Let p_T denote the proportion of individuals given standard treatment plus Infliximab whose condition improves by at least 50% on the ACR scale over a year, and let p_C be the corresponding proportion for individuals given standard therapy plus a placebo. The trial hypotheses can then be written as

$$H_0 : p_T = p_C, \qquad H_1 : p_T \neq p_C.$$

Solution 11.2

(a) The significance level α is

$P(\text{rejecting } H_0, \text{ given that } H_0 \text{ is true})$

$= 1 - P(\text{not rejecting } H_0, \text{ given that } H_0 \text{ is true})$

$= 1 - 0.97 = 0.03.$

The power of the test γ is

$P(\text{rejecting } H_0, \text{ given that } H_0 \text{ is false})$

$= 1 - P(\text{not rejecting } H_0, \text{ given that } H_0 \text{ is false})$

$= 1 - 0.10 = 0.90.$

(b) If a Type I error is made, then the trial will show that the vaccine protects against HIV infection, when in fact it does not. Widespread use of an ineffective vaccine, based on the false premise that the vaccine is effective, may reduce the use of other protective measures (such as safe sex) and result in an increased HIV infection rate.

If a Type II error is made, the trial will fail to show that the vaccine protects against HIV infection, when in fact it does. Since the vaccine will not be widely used, opportunities to prevent HIV infection will be lost.

Solution 11.3

(a) For $\alpha = 0.01$ and $\gamma = 0.8$, we have

$q_{1-\alpha/2} = q_{0.995} = 2.576, \; q_{0.8} = 0.8416.$

Given $\pi_T = 0.2$ and $\pi_C = 0.4$, we have

$\pi_0 = (0.2 + 0.4)/2 = 0.3.$

The sample size required in each group is given by Formula (11.1):

$$n = \frac{2(q_{1-\alpha/2} + q_\gamma)^2 \pi_0(1-\pi_0)}{(\pi_T - \pi_C)^2}$$

$$= \frac{2 \times (2.576 + 0.8416)^2 \times 0.3 \times (1 - 0.3)}{(0.2 - 0.4)^2}$$

$= 122.63\ldots$

$\simeq 123.$

Hence the total sample size required is 246.

(b) For $\alpha = 0.05$ and $\gamma = 0.8$,

$q_{1-\alpha/2} = q_{0.975} = 1.96, \; q_{0.8} = 0.8416.$

Since $\pi_T = 0.6$ and $\pi_C = 0.4$, we have

$\pi_0 = (0.6 + 0.4)/2 = 0.5.$

The sample size required in each group is given by Formula (11.1):

$$n = \frac{2 \times (1.96 + 0.8416)^2 \times 0.5 \times (1 - 0.5)}{(0.6 - 0.4)^2}$$

$= 98.11\ldots$

$\simeq 99.$

Note that the sample size is rounded up to the next integer.

(c) Allowing for 10% losses, the number required for the total sample for the trial in part (a) is $246/0.9 \simeq 274$. The number required for the sample size per group for the trial in part (b) is $99/0.9 = 110$.

Solution 11.4

(a) If $\pi_C = 0.2$, then $\pi_T = 0.2 - 0.1 = 0.1$ and $\pi_0 = (0.1 + 0.2)/2 = 0.15$. The sample size required in each group is given by Formula (11.1):

$$n = \frac{2(q_{1-\alpha/2} + q_\gamma)^2 \pi_0(1-\pi_0)}{(\pi_T - \pi_C)^2}$$

$$= \frac{2 \times (1.96 + 0.8416)^2 \times 0.15 \times (1 - 0.15)}{(0.1)^2}$$

$= 200.14\ldots$

$\simeq 201.$

Hence the total sample size required is 402.

If $\pi_C = 0.4$, then $\pi_T = 0.4 - 0.1 = 0.3$ and $\pi_0 = (0.3 + 0.4)/2 = 0.35$. The sample size required in each group in this case is

$$n = \frac{2 \times (1.96 + 0.8416)^2 \times 0.35 \times (1 - 0.35)}{(0.1)^2}$$

$= 357.12\ldots$

$\simeq 358.$

Hence the total sample size required is 716.

(b) With $\pi_C = 0.2$, the total sample size required is 402, with $\pi_C = 0.4$, it is 716. Thus the sample size required varies considerably for values of π_C between 0.2 and 0.4. If there is genuine uncertainty about the value of π_C then, to be on the safe side, it is best to choose a sample size of 716.

Solution 11.5

(a) Since $\alpha = 0.05$ and $\gamma = 0.9$, we have

$q_{1-\alpha/2} = q_{0.975} = 1.96, \quad q_{0.9} = 1.282.$

Since $\pi_C = 0.25$ and $\pi_T - \pi_C = 0.1$, we have

$\pi_T = 0.25 + 0.10 = 0.35$, and hence

$\pi_0 = (0.25 + 0.35)/2 = 0.3.$

The sample size required in each group is given by Formula (11.1):

$$n = \frac{2(q_{1-\alpha/2} + q_\gamma)^2 \pi_0 (1 - \pi_0)}{(\pi_T - \pi_C)^2}$$

$$= \frac{2 \times (1.96 + 1.282)^2 \times 0.3 \times (1 - 0.3)}{0.1^2}$$

$$= 441.44\ldots$$

$$\simeq 442.$$

(b) For the adverse events, $\pi_C = 0.1$, $\pi_T - \pi_C = 0.05$, so $\pi_T = 0.1 + 0.05 = 0.15$ and hence $\pi_0 = (0.1 + 0.15)/2 = 0.125$.

Thus, using Formula (11.2), for $n = 442$,

$$q_\gamma = |\pi_T - \pi_C| \sqrt{\frac{n}{2\pi_0 (1 - \pi_0)}} - q_{1-\alpha/2}$$

$$= |0.15 - 0.1| \times \sqrt{\frac{442}{2 \times 0.125 \times (1 - 0.125)}} - 1.96$$

$$\simeq 2.248 - 1.96$$

$$\simeq 0.29.$$

Reference to the table of probabilities for the standard normal distribution in the *Handbook* gives $\gamma = 0.6141$. Thus the power to detect such an effect is about 61%.

Solution 12.1

(a) The pooled estimate of the odds ratio is 3.72, with 95% confidence interval (2.29, 6.03), as displayed on Figure 12.2. This indicates a positive association between steroid injections and improvement of osteoarthritis of the knee. This is shown on the forest plot by the diamond, which lies well to the right of the line indicating an odds ratio of 1 (no association).

(b) Since there are six studies, the null distribution of the test statistic is $\chi^2(5)$. From tables, the 0.7-quantile of $\chi^2(5)$ is 6.06 and the 0.8-quantile is 7.29. The observed value of the test statistic is 6.76, so $0.2 < p < 0.3$. Thus there is little evidence of heterogeneity between studies.

(c) The relative weights are indicated by the sizes of the squares representing the odds ratios on the forest plot. Study 6 contributed most to the pooled estimate of the odds ratio, while study 2 contributed least.

Solution 13.2

(a) The NNT for acetazolamide is given by

$$\text{NNT} = \frac{1}{p_C - p_T} = \frac{1}{0.336 - 0.119} \simeq 4.6.$$

Rounded to a whole number, the NNT is 5.

(b) The value reported in the Abstract is 4. Slight (and sometimes not so slight) discrepancies are quite common in published papers. This underlines the need for careful reading of published studies.

Solution 13.3

The answers are to be found in the last sentence of the Introduction.

(a) The disease outcomes of interest are headache and acute mountain sickness, which is earlier described as 'characterised by headache, lightheadedness, fatigue, nausea and insomnia'.

(b) There are three treatments: ginkgo biloba, acetazolamide, and combined ginkgo and acetazolamide.

Solution 13.4

(a) The definition of acute mountain sickness uses the Lake Louise scoring system. In this trial, individuals with a Lake Louise score of 3 or more, who have headache and at least one other symptom, are classified as suffering from acute mountain sickness. Individuals who do not meet this definition are classified as not suffering from acute mountain sickness.

(b) Headache on its own is a secondary end-point.

Solution 13.5

(a) There are four groups in this trial: the placebo group, the ginkgo group, the acetazolamide group, and the combined ginkgo and acetazolamide group.

(b) No stratification was used in the randomization: participants were 'serially enrolled by randomisation number'.

(c) The trial is double-blind.

Solution 13.6

It is stated that the required power is 80%, so $\gamma = 0.8$, and that the design value π_C is 0.57. To check the sample size calculation, the significance level α and the design value π_T are also required.

Solution 13.7

(a) From the flow chart, or from the bottom row of Table 1, the numbers lost to follow-up were as follows: placebo group 32, ginkgo group 33, acetazolamide group 34, combined group 28. These numbers are quite similar.

(b) From row 2 of Table 1, the average age is 36.6 years. From row 5 of Table 1, 259 participants (about 53%) were enrolled at 4358 m. The footnote states that enrolment occurred in the villages of Dingboche (4358 m) and Pheriche (4280 m). Hence the number of participants enrolled at 4280 m, in the village of Pheriche, was $487 - 259 = 228$ — that is, about 47%.

(c) The data presented in Table 1 do not suggest that there were any substantial differences between the groups, at least as far as the listed variables are concerned.

Solution 13.8

(a) No: 614 participants were randomized, but only the 487 followed up were included in the analysis.

(b) The analysis is by intention to treat because those patients who are included in the analysis are analysed according to the group to which they were allocated, not according to the treatment they actually received.

(c) The authors state that 'there were no significant changes [to Table 2] when the table was reproduced without the data from non-compliant participants'. This implies that a per-protocol analysis yielded similar results.

Solution 13.9

(a) The odds ratio for the association between avoidance of acute mountain sickness and use of ginkgo (relative to placebo) is 0.95, with 95% confidence interval (0.56, 1.62). Since the odds ratio is close to 1, and the 95% confidence interval includes 1, there is little evidence that ginkgo has any effect on the incidence of acute mountain sickness.

(b) The odds ratio for *absence* of severe headache is
$$\widehat{OR} = \frac{(124-24) \times 16}{24 \times (119-16)} = \frac{100 \times 16}{24 \times 103} \simeq 0.6472.$$
The estimated standard error is given by
$$\hat{\sigma} = \sqrt{\frac{1}{100} + \frac{1}{24} + \frac{1}{103} + \frac{1}{16}} \simeq 0.3520.$$
Hence the 95% confidence limits are
$$OR^- \simeq 0.6472 \times \exp(-1.96 \times 0.3520) \simeq 0.32,$$
$$OR^+ \simeq 0.6472 \times \exp(1.96 \times 0.3520) \simeq 1.29.$$
Thus the odds ratio is 0.65, with 95% confidence interval (0.32, 1.29), as reported.

Solution 13.10

(a) The odds ratio for the association between treatment with acetazolamide and absence of acute mountain sickness is 3.76, with 95% confidence interval (1.91, 7.39). The odds ratio is much greater than 1, and its 95% confidence interval is located well above 1. This indicates a positive association between treatment with acetazolamide and absence of acute mountain sickness.

(b) In the acetazolamide group, 4 out of 118 suffered severe acute mountain sickness, compared to 22 out of 119 in the placebo group. So
$$p_T = 4/118 \simeq 0.0339, \quad p_C = 22/119 \simeq 0.1849.$$
Hence
$$\text{NNT} = \frac{1}{p_C - p_T} \simeq \frac{1}{0.1849 - 0.0339} \simeq 6.62,$$
or 7 when rounded to an integer.

(c) The odds ratio for avoidance of acute mountain sickness with the combination treatment is 3.04. This is less than the odds ratio for acetazolamide alone, which was 3.76. Thus there is no evidence to suggest that the combination of acetazolamide and ginkgo is better than acetazolamide alone in preventing acute mountain sickness.

Solution 13.11

(a) A variable may be a confounder if it is associated with both the outcome and the exposure (see Subsection 6.4). Ascent rate is known to be associated with acute mountain sickness — the faster the ascent, the less time there is for the body to acclimatize, and hence the greater the chance of developing acute mountain sickness. If the treatment groups are unbalanced with respect to ascent rate, then ascent rate may be a confounder.

(b) One way to test for an association between spending a night at the midpoint and treatment is to undertake a chi-squared test using the data in Table S.14. (The data have been derived from Table 3 of the paper.)

Table S.14 Night spent at climb midpoint by treatment group

Treatment group	Night spent at midpoint Yes	No	Total
Placebo	24	95	119
Acetazolamide	26	92	118
Ginkgo	14	110	124
Combined	31	95	126

This is a 4 × 2 table, and so the null distribution of the test statistic is chi-squared with degrees of freedom $(4-1) \times (2-1) = 3$. (See Subsection 4.2.)

Solution 13.12

(a) The relative risk for prevention of acute mountain sickness is
$$\widehat{RR} = \frac{(118-14)/118}{(119-40)/119} = \frac{104/118}{79/119} \simeq 1.3276.$$
The confidence interval is calculated using the method described in Subsection 2.2. Using (2.4), the estimated standard error is
$$\hat{\sigma} = \sqrt{\frac{1}{104} - \frac{1}{118} + \frac{1}{79} - \frac{1}{119}} \simeq 0.07346.$$
Then, using (2.3), the 95% confidence limits are
$$RR^- \simeq 1.3276 \times \exp(-1.96 \times 0.07346) \simeq 1.15,$$
$$RR^+ \simeq 1.3276 \times \exp(1.96 \times 0.07346) \simeq 1.53.$$
Thus the relative risk is 1.33, with 95% confidence interval (1.15, 1.53).

(b) The meta-analysis gave a pooled relative risk of 1.22 (0.93, 1.59). The two estimates (1.33 and 1.22) are not substantially different, and the confidence intervals largely overlap. Hence it is plausible that the true relative risk is the same in both studies. In this respect it is not quite true to say that the trial results 'clearly counter' the meta-analysis. However, the trial does clearly counter the claim that 'acetazolamide 500 mg does not work'. The lack of evidence of protection in the meta-analysis is most likely the result of the low power of the studies included in it.

Solutions to Exercises

Solution 1.1

(a) The estimates are as follows:
$$\widehat{RR} = \frac{a/n_1}{c/n_2} = \frac{327/542}{76/277} \simeq 2.20,$$

$$\widehat{OR} = \frac{a \times d}{b \times c} = \frac{327 \times 201}{215 \times 76} \simeq 4.02.$$

(b) The relative risk and the odds ratio are both greater than 1. This suggests a positive association between pre-eclampsia or eclampsia during the first pregnancy and hypertension later in life.

Solution 2.1

(a) In Example 2.1, the odds ratio was estimated to be $\widehat{OR} \simeq 2.41$. As the odds ratio is to be used in intermediate calculations, greater accuracy is required. To four decimal places,
$$\widehat{OR} = \frac{893 \times 2783}{5724 \times 180} \simeq 2.4121.$$
To calculate a confidence interval, the estimated standard error $\hat{\sigma}$ is required:
$$\hat{\sigma} = \sqrt{\frac{1}{a} + \frac{1}{b} + \frac{1}{c} + \frac{1}{d}}$$
$$= \sqrt{\frac{1}{893} + \frac{1}{5724} + \frac{1}{180} + \frac{1}{2783}}$$
$$\simeq 0.08491.$$

For a 99% confidence interval, the 0.995-quantile of $N(0,1)$ is required, namely $z = 2.576$. The confidence limits are
$$OR^- = \widehat{OR} \times \exp(-z \times \hat{\sigma})$$
$$\simeq 2.4121 \times \exp(-2.576 \times 0.08491)$$
$$\simeq 1.94,$$
$$OR^+ = \widehat{OR} \times \exp(z \times \hat{\sigma})$$
$$\simeq 2.4121 \times \exp(2.576 \times 0.08491)$$
$$\simeq 3.00.$$
Thus a 99% confidence interval for the odds ratio is $(1.94, 3.00)$.

(b) If deployment to the Gulf was not associated with PTSD then OR would be 1. This is implausible since the 99% confidence interval for OR is $(1.94, 3.00)$ and hence is located well above 1.

Solution 3.1

(a) In a cohort study, the groups would include a group of vaccinated children (the exposed group) and a group of unvaccinated children. Both groups would be followed through time and the numbers of cases of Hib meningitis in the two groups would be counted and compared.

(b) In a case-control study, the two groups would be a group of Hib meningitis cases and a group of children without Hib meningitis. The numbers of children previously vaccinated against Hib in the two groups would then be identified.

(c) Since Hib meningitis is rare, a cohort study would have to be very large. An advantage of a case-control study is that it would not need to be so large.

(d) The odds ratio provides a good approximation to the relative risk when the disease is rare, as is the case here.

Solution 4.1

The expected frequencies under the null hypothesis of no association are in Table S.15.

Table S.15 Wearing a seat belt and sustaining at least moderately severe injury — expected frequencies

Exposure category	Sustained at least moderately severe injury		
	Yes	No	Total
Not wearing a seat belt	10.48	22.52	33
Wearing a seat belt	16.52	35.48	52
Total	27	58	85

The number of children not wearing a seat belt who sustained at least moderately severe injury (14) is greater than expected (10.48).

The value of the test statistic is
$$\chi^2 = \frac{(14 - 10.48)^2}{10.48} + \frac{(19 - 22.52)^2}{22.52}$$
$$+ \frac{(13 - 16.52)^2}{16.52} + \frac{(39 - 35.48)^2}{35.48}$$
$$\simeq 2.83.$$

The null distribution of the test statistic is approximately chi-squared with degrees of freedom $\nu = (2-1)(2-1) = 1$. Since all the expected frequencies are at least 5, the approximation is adequate.

The 0.9-quantile of $\chi^2(1)$ is 2.71 and the 0.95-quantile is 3.84. The observed value of the test statistic lies between these values, so $0.05 < p < 0.1$. Thus, for children aged 4–14, there is weak evidence of an association between not wearing a seat belt and sustaining at least moderately severe injury in the event of a car accident.

Solution 6.1

(a) The stratum-specific odds ratios are as follows:
$$\widehat{OR}_{<1000} = \frac{12 \times 20}{13 \times 10} \simeq 1.85,$$

$$\widehat{OR}_{1000-1499} = \frac{12 \times 83}{30 \times 24} \simeq 1.38,$$

$$\widehat{OR}_{1500-1999} = \frac{7 \times 124}{11 \times 18} \simeq 4.38,$$

$$\widehat{OR}_{2000+} = \frac{15 \times 426}{38 \times 52} \simeq 3.23.$$

(b) The aggregated table is shown below.

Hospital infection	Died	Alive
Yes	46	92
No	104	653

The odds ratio from the aggregated data is given by
$$\widehat{OR} = \frac{46 \times 653}{92 \times 104} \simeq 3.14.$$

(c) The stratum-specific odds ratios are all above 1, as is the odds ratio from the aggregated data. Thus confounding did not reverse the direction of association, though it might have affected the magnitude of the estimated odds ratio.

Solution 7.1

(a) There are four strata, so the null distribution of the test statistic for Tarone's test is $\chi^2(3)$. The 0.8-quantile of $\chi^2(3)$ is 4.64. Since the value of the test statistic is 4.06, $p > 0.2$. Thus there is little evidence that the odds ratios differ between the four strata.

(b) The Mantel–Haenszel odds ratio is given by Formula (7.1):
$$\widehat{OR}_{MH} = \frac{\sum a_i d_i / N_i}{\sum b_i c_i / N_i}$$

$$= \frac{(12 \times 20/55) + \cdots + (15 \times 426/531)}{(13 \times 10/55) + \cdots + (38 \times 52/531)}$$

$$\simeq \frac{28.5071}{12.1546}$$

$$\simeq 2.35.$$

(c) The unadjusted odds ratio is 3.14, while the odds ratio adjusted for birth weight is 2.35. The two values do not differ greatly. This suggests that birth weight is a confounder, but not a major one, for the association between acquisition of an infection in hospital and death. This confounding tends to exaggerate the strength of association.

Solution 7.2

(a) The Mantel–Haenszel estimate of the odds ratio is given by Formula (7.3):
$$\widehat{OR}_{MH} = \frac{f}{g} = \frac{21}{16} = 1.3125 \simeq 1.31.$$

(b) For a 95% confidence interval, $z = 1.96$. The estimated standard error $\hat{\sigma}$ is given by Formula (7.4):
$$\hat{\sigma} = \sqrt{\frac{1}{f} + \frac{1}{g}} = \sqrt{\frac{1}{21} + \frac{1}{16}} \simeq 0.3318.$$

The 95% confidence limits are given by
$$\widehat{OR}_{MH} \times \exp(\pm 1.96 \times \hat{\sigma}) \simeq 1.3125 \times \exp(\pm 0.6503).$$

Thus an approximate 95% confidence interval for OR is
$$(OR^-, OR^+) \simeq (0.68, 2.51).$$

(c) The value of McNemar's test statistic is found using (7.5):
$$\chi^2 = \frac{(|f - g| - 1)^2}{f + g} = \frac{(|21 - 16| - 1)^2}{21 + 16} \simeq 0.4324.$$

The 0.8-quantile of $\chi^2(1)$ is 1.64, so $p > 0.2$.

(d) The odds ratio is 1.31, indicating a rather weak association between recent infection and stroke. The 95% confidence interval is $(0.68, 2.51)$, which contains 1. McNemar's test provides little evidence $(p > 0.2)$ against the null hypothesis of no association. Overall, the study provides little evidence that recent infection with *Chlamydia pneumoniae* and stroke are associated.

Solution 7.3

(a) The estimated odds ratio for non-smokers is
$$\widehat{OR} = \frac{8 \times 46}{28 \times 3} \simeq 4.38.$$

For smokers, the estimated odds ratio is
$$\widehat{OR} = \frac{40 \times 93}{47 \times 56} \simeq 1.41.$$

There appears to be a large difference between the odds ratios in the two strata. Since the Mantel–Haenszel method is based on the assumption of a common odds ratio, the method may not be applicable in this instance.

(b) There are two strata, hence the null distribution of Tarone's test statistic is $\chi^2(1)$. The observed test statistic is 2.27. The 0.9-quantile of $\chi^2(1)$ is 2.71, and the 0.8-quantile is 1.64. Hence $0.1 < p < 0.2$. Thus there is little evidence against the null hypothesis that the odds ratio is the same for both strata.

(c) In part (a), the estimated odds ratios were rather different, whereas the test in part (b) provided little evidence against the null hypothesis of a common odds ratio. These two findings are not contradictory: since the numbers of non-smokers who died of lung cancer are small, the confidence interval for the odds ratio in non-smokers is very wide. You may conclude from this that it is not unreasonable to estimate a common odds ratio using the Mantel–Haenszel method.

Solutions to Exercises

Solution 8.1

(a) The lowest dose is $< 0.25\,\mathrm{mg/m^3\,yr}$. The odds ratio for the ≥ 100 group relative to the group experiencing the lowest dose is
$$\widehat{OR}_{\geq 100} = \frac{12 \times 35}{9 \times 14} \simeq 3.33.$$
Similarly,
$$\widehat{OR}_{15-100-} = \frac{36 \times 35}{66 \times 14} \simeq 1.36,$$
$$\widehat{OR}_{0.25-15-} = \frac{45 \times 35}{104 \times 14} \simeq 1.08.$$
The dose-specific odds ratios suggest that the risk of dying from lung cancer may increase with dose, though the effect may only be apparent at very high doses.

(b) The null distribution of the test statistic is approximately $\chi^2(1)$. The 0.9-quantile of $\chi^2(1)$ is 2.71 and the 0.95-quantile is 3.84, so $0.05 < p < 0.10$. This provides weak evidence of a linear dose-response relationship between arsenic dose and death from lung cancer.

Solution 10.1

It is required to maintain balance according to age (in two groups: 12–24 months and 25+ months) and previous incidence of otitis media (categorized as low and high). One way to achieve this is by stratified randomization, by blocks within each of four strata: aged 12–24 months with low incidence, aged 12–24 months with high incidence, aged 25–84 months with low incidence, and aged 25–84 months with high incidence.

Solution 10.2

(a) The intention-to-treat analysis included 192 children randomized to the control group (the missing one is the child who was lost to follow-up), and all 190 children randomized to the treatment group. The data table for the analysis is shown below.

Trial group	1+ episodes	No episodes	Total
Treatment	107	83	190
Control	101	91	192

Note that the children included in the intention-to-treat analysis include those who discontinued treatment (but for whom the outcome could be assessed).

(b) The estimated odds ratio is given by
$$\widehat{OR} = \frac{107 \times 91}{83 \times 101} \simeq 1.1615.$$
The estimated standard error $\hat{\sigma}$ is given by
$$\hat{\sigma} = \sqrt{\frac{1}{107} + \frac{1}{83} + \frac{1}{101} + \frac{1}{91}} \simeq 0.2056.$$
Hence the 95% confidence limits are
$$OR^- \simeq 1.1615 \times \exp(-1.96 \times 0.2056) \simeq 0.78,$$
$$OR^+ \simeq 1.1615 \times \exp(1.96 \times 0.2056) \simeq 1.74.$$

(c) The odds ratio is 1.16, with 95% confidence interval $(0.78, 1.74)$. The confidence interval includes values less than 1 (indicating that the vaccine may protect against otitis media) and values greater than 1 (indicating that it may increase the risk of otitis media). Thus the trial provides little evidence that vaccination reduces the number of children who experience otitis media.

Solution 11.1

(a) For the pneumococcal vaccine trial, $\alpha = 0.05$ and $\gamma = 0.8$, so $q_{1-\alpha/2} = q_{0.975} = 1.96$ and $q_{0.8} = 0.8416$. Also $\pi_C = 0.55$ and $\pi_T = 0.4$, so
$$\pi_0 = (0.55 + 0.4)/2 = 0.475.$$
The sample size required per group is given by Formula (11.1):
$$n = \frac{2(q_{1-\alpha/2} + q_\gamma)^2 \pi_0 (1-\pi_0)}{(\pi_T - \pi_C)^2}$$
$$= \frac{2 \times (1.96 + 0.8416)^2 \times 0.475 \times (1-0.475)}{(0.4 - 0.55)^2}$$
$$= 173.98\ldots$$
$$\simeq 174.$$

(b) Allowing for 10% losses, the number required per group is $174/0.9 \simeq 194$. Thus the total sample size to be randomized is $2 \times 194 = 388$. In the actual trial, 383 were randomized.

Solution 12.1

(a) Five studies supported a positive association, insofar as their odds ratios were greater than 1. Only study 1 supported a negative association.

(b) Since there are six studies, the null distribution of the test statistic is $\chi^2(5)$. From tables, the 0.8-quantile of $\chi^2(5)$ is 7.29 and the 0.9-quantile is 9.24. The observed value of the test statistic is 7.31, so $0.1 < p < 0.2$. There is little evidence of heterogeneity between studies.

(c) The pooled estimate of the odds ratio is 1.25, with 95% confidence interval $(1.13, 1.40)$. This suggests that HRT is positively associated with breast cancer.

Solution 14.1

This exercise covers some of the ideas and techniques discussed in Sections 1 and 2.

(a) The estimated relative risk is given by Formula (1.1):
$$\widehat{RR} = \frac{a/n_1}{c/n_2} = \frac{40/3420}{41/10\,588} \simeq 3.0204.$$

(b) To calculate a confidence interval, the estimated standard error is required. Using Formula (2.4),
$$\widehat{\sigma} = \sqrt{\frac{1}{a} - \frac{1}{n_1} + \frac{1}{c} - \frac{1}{n_2}}$$
$$= \sqrt{\frac{1}{40} - \frac{1}{3420} + \frac{1}{41} - \frac{1}{10\,588}}$$
$$\simeq 0.2214.$$

For a 95% confidence interval, $z = 1.96$. The 95% confidence limits are
$$RR^- = \widehat{RR} \times \exp(-z \times \widehat{\sigma})$$
$$\simeq 3.0204 \times \exp(-1.96 \times 0.2214)$$
$$\simeq 1.96,$$
$$RR^+ = \widehat{RR} \times \exp(+z \times \widehat{\sigma})$$
$$\simeq 3.0204 \times \exp(1.96 \times 0.2214)$$
$$\simeq 4.66.$$

(c) The relative risk is 3.02, with 95% confidence interval $(1.96, 4.66)$. The confidence interval lies well above 1. Acquiring an *S. aureus* blood infection in hospital is positively associated with nasal carriage on admission.

Solution 14.2

This exercise covers some of the ideas and techniques discussed in Section 4.

(a) Using (4.1), the expected frequencies under the null hypothesis of no association are as shown in Table S.16.

Table S.16 Expected frequencies for asthma in Chinese cities

City	Asthma Yes	Asthma No	Total
Hong Kong	130.94	2 979.06	3 110
Guangzhou	150.09	3 414.91	3 565
Beijing	177.97	4 049.03	4 227
Total	459	10 443	10 902

The observed value of the chi-squared test statistic is
$$\chi^2 = \frac{(179 - 130.94)^2}{130.94} + \frac{(121 - 150.09)^2}{150.09}$$
$$+ \frac{(159 - 177.97)^2}{177.97} + \frac{(2931 - 2979.06)^2}{2979.06}$$
$$+ \frac{(3444 - 3414.91)^2}{3414.91} + \frac{(4068 - 4049.03)^2}{4049.03}$$
$$\simeq 26.41.$$

(b) Under the null hypothesis, the distribution of the chi-squared test statistic is approximately chi-squared with degrees of freedom $\nu = (3-1) \times (2-1) = 2$. Since all the expected frequencies are greater than 5, the approximation is adequate. From tables, the 0.999-quantile of $\chi^2(2)$ is 13.82. Since the observed value of the test statistic is greater than this, $p < 0.001$.

(c) There is strong evidence against the null hypothesis that the proportion of children with asthma is not associated with city. The numbers of asthma sufferers are higher than expected in Hong Kong, and lower than expected in Guangzhou and Beijing.

Solution 14.3

This exercise covers some of the ideas and techniques discussed in Sections 2, 3 and 6.

(a) The estimated odds ratio is
$$\widehat{OR} = \frac{a \times d}{b \times c} = \frac{49 \times 985}{242 \times 408} \simeq 0.4888.$$

The estimated standard error $\widehat{\sigma}$ is given by
$$\widehat{\sigma} = \sqrt{\frac{1}{a} + \frac{1}{b} + \frac{1}{c} + \frac{1}{d}}$$
$$= \sqrt{\frac{1}{49} + \frac{1}{242} + \frac{1}{408} + \frac{1}{985}}$$
$$\simeq 0.1674.$$

For a 95% confidence interval, $z = 1.96$. Thus the 95% confidence limits are
$$OR^- = \widehat{OR} \times \exp(-z \times \widehat{\sigma})$$
$$\simeq 0.4888 \times \exp(-1.96 \times 0.1674)$$
$$\simeq 0.35,$$
$$OR^+ = \widehat{OR} \times \exp(z \times \widehat{\sigma})$$
$$\simeq 0.4888 \times \exp(1.96 \times 0.1674)$$
$$\simeq 0.68.$$

So the odds ratio is 0.49, with 95% confidence interval $(0.35, 0.68)$.

(b) The odds ratio is less than 1, and the confidence interval is located entirely below 1. This suggests a negative association between wearing high-visibility clothing and sustaining motorcycle crash injuries: wearing such clothing reduces the risk of injury.

(c) Age could confound the association if it were associated both with the risk of injury, and with the wearing of high-visibility clothing. However, adjustment for the effect of age produces little change in the odds ratio or its confidence interval. Thus confounding by age did not occur in this study.

Solution 14.4

This exercise covers some of the ideas and techniques discussed in Section 7.

(a) The Mantel–Haenszel estimate of the odds ratio is obtained using Formula (7.3):
$$\widehat{OR}_{MH} = \frac{f}{g} = \frac{59}{99} \simeq 0.5960.$$
The estimated standard error $\hat{\sigma}$ is obtained using (7.4):
$$\hat{\sigma} = \sqrt{\frac{1}{f} + \frac{1}{g}} = \sqrt{\frac{1}{59} + \frac{1}{99}} \simeq 0.1645.$$
For a 95% confidence interval $z = 1.96$. Thus the 95% confidence limits are
$$OR^- = \widehat{OR}_{MH} \times \exp(-z \times \hat{\sigma})$$
$$\simeq 0.5960 \times \exp(-1.96 \times 0.1645)$$
$$\simeq 0.43,$$
$$OR^+ = \widehat{OR}_{MH} \times \exp(z \times \hat{\sigma})$$
$$\simeq 0.5960 \times \exp(1.96 \times 0.1645)$$
$$\simeq 0.82.$$
So the odds ratio is 0.60, with 95% confidence interval $(0.43, 0.82)$.

(b) The value of McNemar's test statistic is
$$\chi^2 = \frac{(|f - g| - 1)^2}{f + g} = \frac{(|59 - 99| - 1)^2}{59 + 99} \simeq 9.63.$$
Under the null hypothesis of no association, the distribution of the test statistic is approximately $\chi^2(1)$. The 0.995-quantile of $\chi^2(1)$ is 7.88 and the 0.999-quantile is 10.83. Hence $0.001 < p < 0.005$. There is strong evidence of association between taking physical exercise and suffering a heart attack.

(c) The odds ratio is 0.60 and the confidence interval is located entirely below 1, which suggests a negative association. The p value for McNemar's test is less than 0.005. This indicates that there is strong evidence of association between taking physical exercise and suffering a heart attack. Together these results indicate that taking physical exercise is associated with a reduction in the risk of heart attack.

Solution 14.5

This exercise covers some of the ideas and techniques discussed in Section 8.

(a) The dose-specific odds ratios relative to the Never category are as follows:
$$\widehat{OR}_{\text{Always}} = \frac{499 \times 1504}{316 \times 993} \simeq 2.39,$$
$$\widehat{OR}_{\text{Often}} = \frac{1125 \times 1504}{1117 \times 993} \simeq 1.53,$$
$$\widehat{OR}_{\text{Sometimes}} = \frac{2265 \times 1504}{3315 \times 993} \simeq 1.03.$$
The dose-specific odds ratios increase with the level of exposure to stress. This suggests that there may be a positive dose-response relationship in the association between stress at work and myocardial infarction.

(b) Under the null hypothesis of no linear dose-response relationship, the distribution of the test statistic is approximately $\chi^2(1)$. The 0.999-quantile of $\chi^2(1)$ is 10.83. The observed value is much greater than this, so $p < 0.001$. There is strong evidence of a dose-response relationship between stress at work and myocardial infarction in males, and from part (a) the relationship is positive: the risk of myocardial infarction increases with stress levels.

Solution 14.6

This exercise covers some of the ideas and techniques discussed in Sections 10 and 11.

(a) Randomization by blocks of 20 means that, in each successive group of 20 patients, ten were assigned randomly to the treatment group and ten to the control group.

(b) This method of analysis is called analysis by intention to treat. Its rationale is to reduce the impact on the results of selection biases that might arise after randomization, through protocol violations (for example, patients switching treatments).

(c) For $\pi_C = 0.15$ and $\pi_T - \pi_C = -0.10$, we have $\pi_T = 0.05$, so
$$\pi_0 = (\pi_T + \pi_C)/2 = (0.05 + 0.15)/2 = 0.10.$$
For 80% power and 5% significance level, $q_{1-\alpha/2} = 1.96$ and $q_\gamma = 0.8416$. Hence, using Formula (11.1), the sample size required in each group is
$$n = \frac{2(q_{1-\alpha/2} + q_\gamma)^2 \pi_0(1 - \pi_0)}{(\pi_T - \pi_C)^2}$$
$$= \frac{2 \times (1.96 + 0.8416)^2 \times 0.10 \times (1 - 0.10)}{(-0.10)^2}$$
$$= 141.28\ldots.$$
Thus 142 patients are required in each group.

Solution 14.7

This exercise covers some of the ideas and techniques discussed in Section 12.

(a) Since autism is an uncommon condition, relative risks and odds ratios are virtually identical (see Subsection 1.3). Thus they can be compared and pooled in a meta-analysis.

(b) Study 4 has the greatest weight, study 3 the smallest, as indicated by the sizes of the squares.

(c) The four studies separately provide little evidence of an association between MMR vaccination and autism. All report odds ratios or relative risks less than 1. The pooled effect is 0.87, with 95% confidence interval $(0.76, 1.00)$. This does not suggest that MMR vaccination is positively associated with autism; on the contrary, it provides some (rather weak) evidence that MMR vaccination may protect against autism.

Index

aggregated table 49
association
    measures of 14
    negative 9, 11
    positive 9, 11
    strength of 12, 13

balance 74
bias 42
binomial model 17
blinding 76
block size 75

case-control study 23, 24
    1–1 matched 57, 62
    analysis of 1–1 matched 62
causality 7
causation 42
causation criteria 66
chi-squared distribution 35
chi-squared test 34
    for no association 35, 39
    for no linear trend 69
chi-squared test statistic 34
clinical trials 82
cohort 8
cohort study 7, 8
    controlled 9
    retrospective 11
combining evidence 94
confidence interval 18
    for the Mantel–Haenszel odds ratio 60, 62
    for the odds ratio 21
    for the relative risk 19
confounder 49, 51
confounding 42, 43, 47
confounding bias 51
control group 8
controls 23
crude odds ratio 55

Data Monitoring Committee 82
degrees of freedom 35
design values 88
discordant pairs 58
dose 67
dose-response relationship 67
dose-specific odds ratio 67
double-blind 76

effect modifier 63
epidemiology 6
error probabilities 87
ethical approval 82
evaluation of pharmaceutical drugs 81
exact test 40
expected frequency 33
exposed group 8
exposure 8

Fisher, R.A. 40
Fisher's exact test 40
fixed-level testing 85
flow chart 80
forest plot 96

heterogeneous 95
Hill, A.B. 66
Hill's criteria for causation 66
hypotheses 85
hypothesis test 85

information bias 43, 46
intention to treat 79
interaction 62
interpretation of p values 38
intervention group 73

levels 49

Mantel–Haenszel odds ratio 53–55
marginal totals 33
matched case-control set 57
matching 57
McNemar's test 60
measures of association 14
    in case-control studies 27
medical literature 100
meta-analysis 72, 95
multi-centre trials 75

number needed to harm 106
number needed to treat 105

odds 13
odds ratio 13
    Mantel–Haenszel 53–55
    pooled estimate 95
    stratum-specific 49
    unadjusted 55
open trial 77
outcome 8

Pearson, K. 34
per protocol 79
phases of pharmaceutical drug evaluation 82
power 87
    for a given sample size 92
practically significant difference 91
primary outcome 93
protocol 73
protocol violations 79
publication bias 94

randomization 73
randomize by blocks 75
randomized controlled trials 72
reference exposure category 29
relative risk 11

Index

risk   11
risk factor   7

sample size   85, 89
secondary outcomes   93
selection bias   43
sensitivity analysis   91
significance level   85, 87
Simpson's paradox   49
single-blind   78
standard error   18
strata   49
stratified randomization   75
stratifying variable   49, 62
stratum-specific odds ratio   49
strength of association   39
strength of evidence   39
systematic review   94

Tarone's test for homogeneity   64
treatment group   73
trial protocol   73
Type I error   87
Type II error   87

unadjusted odds ratio   55